Y0-BSD-662

Computer-Assisted Organic Synthesis

W. Todd Wipke, EDITOR
University of California, Santa Cruz

W. Jeffrey Howe, EDITOR
The Upjohn Company

A symposium cosponsored by the Division of Chemical Information and the Division of Computers in Chemistry at the Centennial Meeting of the American Chemical Society, New York, N.Y., April 7–8, 1976.

ACS SYMPOSIUM SERIES **61**

AMERICAN CHEMICAL SOCIETY
WASHINGTON, D. C. 1977

SEP
6212-1765 ✓

CHEM

Library of Congress CIP Data

Computer-assisted organic synthesis.
(ACS symposium series; 61 ISSN 0097-6156)

Includes bibliographical references and index.

1. Chemistry, Organic—Synthesis—Data processing—Congresses.
I. Wipke, W. Todd, 1940- . II. Howe, William Jeffrey, 1946- . III. American Chemical Society. Division of Chemical Information. IV. American Chemical Society. Division of Computers in Chemistry. V. American Chemical Society. VI. Series: American Chemical Society. ACS symposium series; 61.

QD262.C54 547′.2′0285 77-13629
ISBN 0-8412-0394-6 ACSMC8 61 1-239

PRINTED IN THE UNITED STATES OF AMERICA

FOREWORD

The ACS SYMPOSIUM SERIES was founded in 1974 to provide a medium for publishing symposia quickly in book form. The format of the SERIES parallels that of the continuing ADVANCES IN CHEMISTRY SERIES except that in order to save time the papers are not typeset but are reproduced as they are submitted by the authors in camera-ready form. As a further means of saving time, the papers are not edited or reviewed except by the symposium chairman, who becomes editor of the book. Papers published in the ACS SYMPOSIUM SERIES are original contributions not published elsewhere in whole or major part and include reports of research as well as reviews since symposia may embrace both types of presentation.

CONTENTS

PREFACE

The application of digital computers to the design and study of organic syntheses has stimulated the interest of chemists and computer scientists alike. The field not only promises practical benefits in synthesis planning and chemical education but also contributes to many areas of theoretical interest such as synthetic strategies, reaction indexing, structure representation, substructural perception, computer graphics, molecular modelling, and artificial intelligence. Many concepts and techniques now being used in structure elucidation, structure–activity, and other chemical information systems were first developed in the computer synthesis planning field.

Since 1969 when the first paper in this area appeared, several groups have pursued approaches to computer synthesis planning. The symposium from which this volume derives was the first meeting of all the major research groups in this field. The papers in this volume describe the state of the art of computer synthesis as viewed by the major research groups working in the area.

The editors acknowledge the Petroleum Research Fund for partial support of the symposium through a travel grant to Ivar Ugi and thank Cynthia O'Donohue for her help as program chairman.

W. Todd Wipke
Santa Cruz, CA
August 1977

W. Jeffrey Howe
Kalamazoo, MI

1

LHASA—Logic and Heuristics Applied to Synthetic Analysis

DAVID A. PENSAK

Central Research and Develop. Dept., E. I. du Pont de Nemours and Co., Wilmington, Del. 19898

E. J. COREY

Dept. of Chemistry, Harvard University, Cambridge, Mass. 02138

Despite the wealth of knowledge about various chemical reactions, there exists no formal framework of interrelationships to guide the chemist in the synthesis of even moderately complex molecules. The LHASA (Logic and Heuristics Applied to Synthetic Analysis) project is an attempt to codify and organize the techniques used in organic synthesis.

One important aspect of the project has been the writing of a general purpose computer program which will aid the laboratory chemist and will employ both the basic and more complex techniques for synthetic design as elucidated by this study. The program (hereafter also called LHASA) is intended to propose a variety of synthetic routes to whatever molecule it is given. The responsibility for final evaluation of the merit of the routes lies with the chemist. The program is to be an adjunct to the laboratory chemist as much as any analytical tool.

Since LHASA is incapable of proposing any routes the chemist could not have thought of by himself, i.e., it does not suggest new reactions that have never been tried, there needs to be some justificationfor the massive effort involved in writing the program. It is well known that humanity and creativity bring with them certain unavoidable shortsightedness and prejudices. There will be particular reactions with which each chemist is most familiar and it is to these that he will first look to find his synthetic route(s). It is precisely this failure to consider all possible routes that makes a program like LHASA both useful and necessary. Computers are well known for their ability to perform rote tasks a great number of times without complaint. Examination of potential synthetic pathways may be broken down into sufficiently small steps as to be amenable to computer implementation.

How should one go about designing a synthesis? One of the most basic techniques is to work backwards, the final product or *target molecule* being the ultimate goal. This target is examined to find any or all compounds which can be transformed into it in a single chemical step. Each of these precursors may then be similarly analyzed until satisfactory starting materials are obtained. This method of analysis is called *retrosynthetic* (or, equivalently, *antithetic*). A structural modification that is being performed in the retrosynthetic direction is called a *transform* and is graphically depicted as a double arrow.

OAc, CH_3, CH_2R ⇌ OH, CH_3, CH_2R ⇒ O, CH_3 + RCH_2Br

When applied in its most general form, retrosynthetic analysis could be applied to every precursor of the target molecule and then, in turn, to each of the new structures. The graphical representation of such an analysis is called a *synthesis tree*. (A complex example is shown below). It is worthwhile to note that structures tend to be less complex the further away they are from the target molecule and no constraints are placed on the choice of reactions. The starting materials are the last to be generated, thereby maintaining flexibility in the choice of route until the end of the analysis.

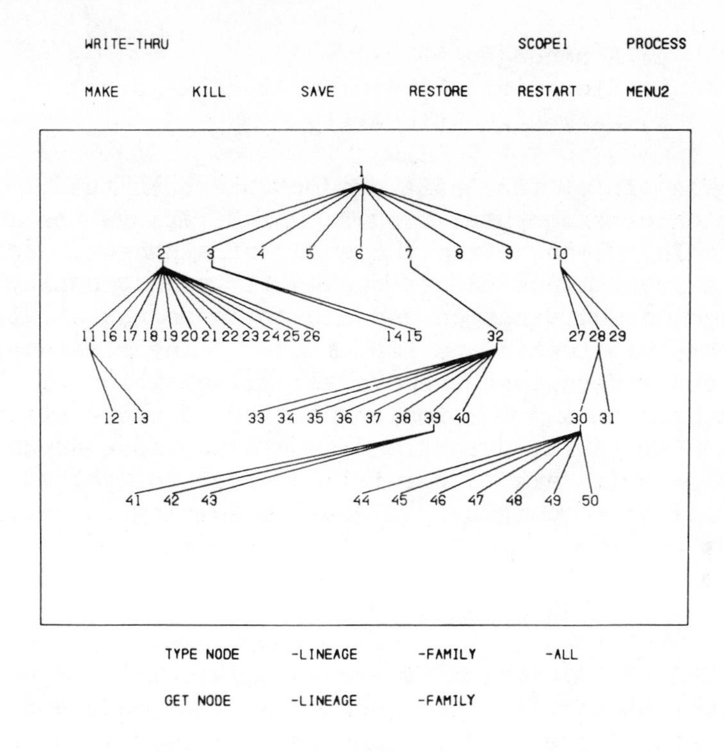

Of considerable importance to the way a synthesis is analyzed is the detailed plan of execution. The ordering of the addition (or removal) of individual functionalities, the manipulation of stereocenters, and the closure of rings can be of crucial importance in terms of interferences or competing reactions. The procedures for choosing the sequence in which these discrete steps are performed are called strategies.

At present there are three broad categories of synthetic strategy that LHASA is capable of employing. They are

1) Opportunistic or Functional Group Based Strategies
2) Strategic Bond Disconnections for Polycyclic Targets
3) Strategies Based on Structural Features

a) Appendages
b) Rings (small, common, medium-large)
c) Masked Functionality

The aim of the LHASA project has been the creation of a computer program which employs the strategies gleaned from the study of synthetic design. Such a program now exists though it is continually undergoing modification and expansion as new strategies are elucidated and implemented. The remainder of this paper describes the overall organization of the LHASA program and the implementation of these strategies. Particular attention is paid to those aspects of LHASA which are of special interest to synthetic chemists or to computer scientists working in chemical areas.

ORGANIZATION OF LHASA

The LHASA program is exceedingly complex - about 400 subroutines, 30,000 lines of FORTRAN code and a data base of over 600 common chemical reactions. To describe it in detail is well beyond the scope of this paper.[1] A general overview is relevant as it puts the functions of the data base in a reasonable perspective. Figure 1 shows a global view of LHASA.

---

[1]See for example
Corey, E. J., W. J. Howe and D. A. Pensak, J. Amer. Chem. Soc., 96, 7724 (1974).
Corey, E. J., Quart. Rev. Chem. Soc., 25, 455 (1971).
Corey, E. J., W. T. Wipke, R. D. Cramer III and W. J. Howe, J. Amer. Chem. Soc., 94, 421 (1972).
Corey, E. J., W. L. Jorgensen, J. Amer. Chem. Soc., 98, 189 (1976).

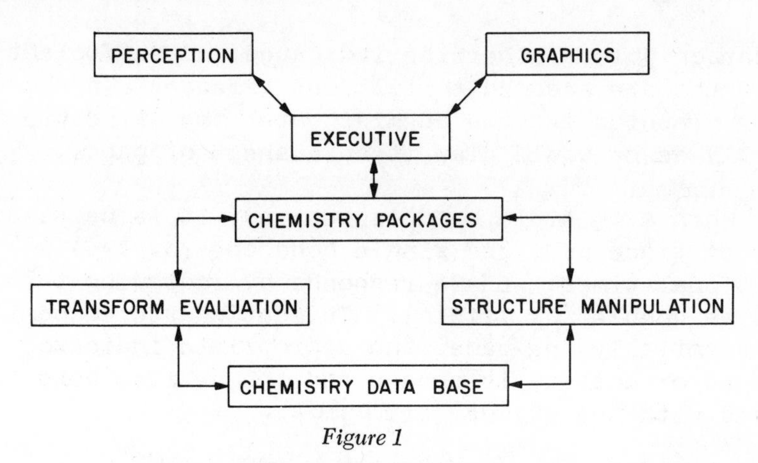

Figure 1

Sample recognition questions

GRAPHICS

There is no doubt that the part of LHASA which makes the most immediate impact on chemists is the graphics. He may draw in the structure that he is interested in analyzing using standard chemical conventions and all communication from the program to him is via structural diagrams. This manner of interfacing with the chemist-user was chosen because it has been shown that the rate at which he can assimilate chemical data is maximized if it is in the notation with which he is most familiar.

As the chemist sits facing the CRT (cathode ray tube) with which he communicates with LHASA, on his right is a digitizing data tablet. This is a device which measures the two dimensional coordinates of the stylus or pen which is used for inputting structures. As the stylus is moved along the surface of the tablet, LHASA is told by the tablet where the stylus touched the surface and where it had moved to when it was lifted. A line is displayed on the CRT between these two points exactly like a bond being drawn on a sheet of paper. The LHASA graphics routines recognize that two atoms are required to make this bond and makes appropriate internal entries in the program data base. Conforming to standard structure conventions, an atom is carbon unless otherwise indicated and sufficient hydrogens are assumed to fill out valence. In short, LHASA graphics let the chemist enter the structure exactly as he would draw it on a sheet of paper.

When a multiple bond is desired, it is necessary only to trace over the single bond one (or two) additional times. LHASA responds by redrawing the bond as double (or triple). Indicating stereochemistry is essentially the same, the appropriate indicator (wedged or dotted) is chosen and the desired bond is traced with the stylus (see below).

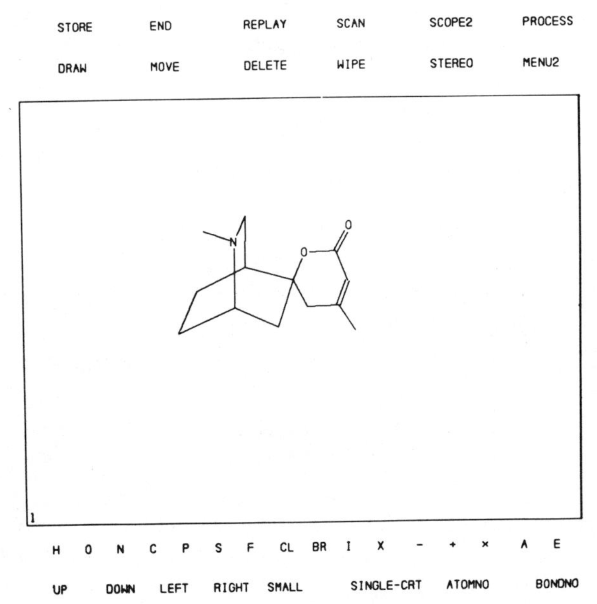

Considerable effort was expended to insure that no artistic talent is required in structural input. A reasonable amount of inaccuracy is permitted in pointing to an atom or bond. LHASA determines what was intended and acts accordingly. There is no need to worry about consistency or precision of bonds or angles. With the one exception of indicating cis-trans isomerism around double bonds, it makes no difference how distorted the structure is input. LHASA works solely from information about connectivity. This flexibility is quite useful for drawing structures in a particular conformation. The program will process it correctly and display all offspring in the same conformation.

The only other times that the chemist must communicate with LHASA are when decisions are to be made about which structure is to be analyzed and what method is to be used. In all cases a table of choices (or menu) is displayed and the chemist merely points to the one (or ones) that he wishes. A sample menu is shown below.

<u>SINGLE GROUP</u>

FULL SEARCH

BOND MODE

SUBGOALS

SEQUENTIAL FGI

ISOLATED STRAT BOND

DISCONNECTIVE

RECONNECTIVE

UNMASKING

<u>GROUP PAIR</u>

FULL SEARCH

BOND MODE

NARROW MODE

MANUAL MODE

APPENDAGE CHEMISTRY

RING APPNDG ONLY

BRANCH APPNDG ONLY

PERCEPTION ONLY

DEBUG

TREE

EXIT

<u>KEY SUBSTRUCTURES</u>

DIELS-ALDER

ROBINSON 4+2

ROBINSON 3+3

SMALL RINGS

STEREOSPECIFIC C=C

At no time is he forced to learn any special command formats or memorize lists of options. The LHASA graphics and control structures were specifically designed to be as easy and natural for the chemist to use as possible.

<u>PERCEPTION</u>

Inherent in a structural diagram is a wealth of information - rings, functional groups, stereochemistry, etc. To be as effective as possible, LHASA must recognize all of these and utilize them in its planning processes. The procedures by which these are called <u>perception</u>.

By virtue of having very sophisticated perception routines, LHASA can avoid forcing the chemist to input any artificial (to him) information. An unexpected adjunct of this has been a guarantee of perceptual completeness. For example, consider the following structure (the non-indole portion of the alkaloid ajmaline).

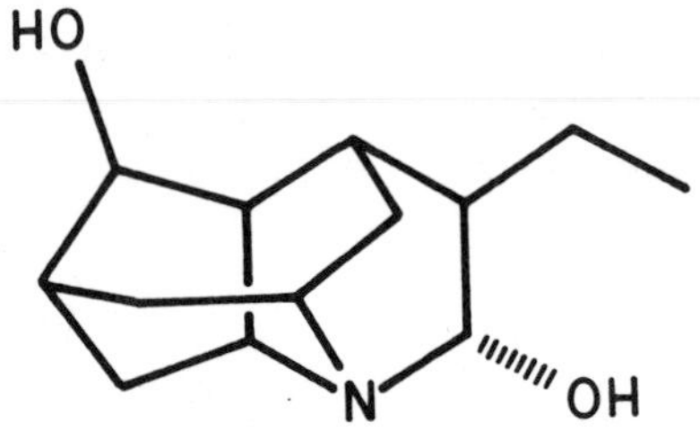

HO

N OH

There are three six-membered rings, one five-membered ring and one seven-membered ring yet few chemists perceived all of them. If the structure were redrawn as

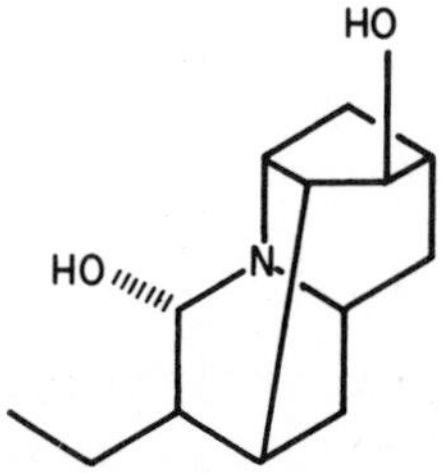

HO

HO N

a different, though still incomplete set of rings is recognized by the human. The point of this example is that LHASA must perceive rings solely on the basis of connectivity, not how the structure is drawn. The program would be useless if it missed syntheses based on cyclic substructures because the chemist had failed to indicate all the rings in the molecule.

RING PERCEPTION

Many researchers have attacked the problem of finding the set of cycles in a graph.[2] Their work has primarily been directed towards identifying the smallest set rings of rings in the network. For chemical purposes, this is not sufficient however. For example, the structure below

is best synthesized by the Diels Alder addition shown, but the six-membered ring formed is not part of the minimal cyclic basis of the molecule. It is necessary, therefore, to redefine our problem as that of finding the set of cycles in a graph which are of chemical significance.

For synthetic purposes, rings must be split into two classes - <u>real</u> and <u>pseudo</u>. For each bond in a molecule, the smallest ring containing that bond is called a real ring. It is quite possible that the number of real rings will be greater than the cyclic order of the molecule (# bonds - # atoms + 1). Pseudo rings are the pairwise envelopes of real rings with the restriction that the size of the envelope be seven or less. Real rings are useful because the chemistry of a bond is best reflected by the size of the smallest ring containing it - for example, the fusion bond in the structure below.

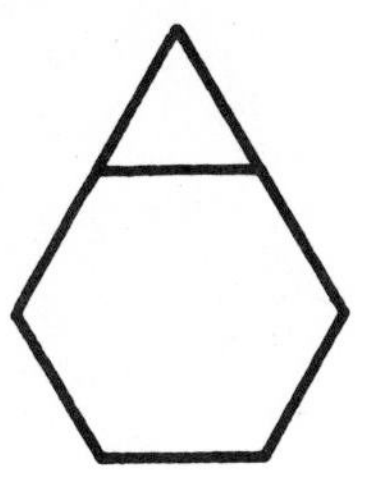

[2] See for example
Paton, K., *Commun. Assoc. Comput. Mach.*, 12, 514 (1969).

Pseudo rings are useful because they are the rings which are often formed in the construction of bridged molecules.

STRATEGIC BONDS - CYCLIC

There are usually certain bonds in a molecule whose disconnection in the retrosynthetic direction leads to a significant simplification of the cyclic structure. These are termed strategic bonds. Since these have been described in detail elsewhere, we shall consider them here only briefly.[3]

The first premise of strategic bonds is that the chemical activity of a bond is a direct function of the size of the smallest ring containing it. This leads to the requirement that a strategic bond must be in a ring of five, six, or seven members and not in or exo to a cyclopropyl ring. A strategic bond must also be in the ring (if any) with the maximum number of bridges on it. This insures that its disconnection will simplify the cyclic network as much as possible. The structure below shows the power of this heuristic.

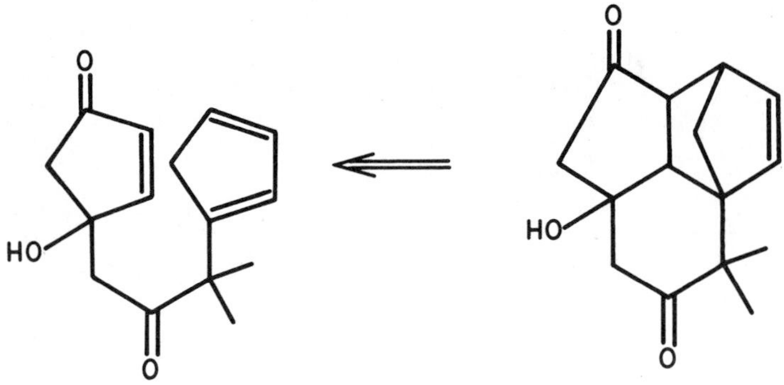

[3] Corey, E. J., W. J. Howe, H. W. Orf, D. A. Pensak and G. A. Petersson, J. Amer. Chem. Soc., 97, 6116 (1975).

Other restrictions on strategic bonds prevent them from being aromatic and from leaving chiral side chains. These requirements are also based on current chemical technique.

FUNCTIONAL GROUP PERCEPTION

Functionality in molecules is well defined. There is no argument about whether a group is a ketone or not. The problem in LHASA is to perceive all groups and juxtapositions of groups which can be chemically meaningful.

To this end, a context dependent grammar has been developed to unambiguously represent the physical domain of a group and the site(s) at which it can be expected to react. This grammar defines which atoms in the group are considered origins. Each functional group is characterized by at least one carbon atom which is its attachment to the rest of the molecule. This is called an origin atom and it is around these origins that many of the data tables in LHASA are organized. As an example, an alpha-beta unsaturated ketone has two olefinic positions with significantly different affinity to electrophilic reagents. To consider the double bond as one group with constant reactivity is an unreasonable simplification, but recognizing it as two origin atoms each with olefinic character makes suitable differentiation possible.

To accomplish this recognition efficiently, a table driven approach was chosen since the types of groups to be recognized change from time to time, depending on the needs of those using the program (sixty-four different groups are currently recognized by LHASA, see Table 1). This consists of a series of questions which can be answered with either a "yes" or "no". Depending on the answer, the type of the next question to be asked is specifically determined. For

Table 1

KETONE	SULFONE
ALDEHYDE	C*SULFONATE
ACID	LACTAM
ESTER	PHOSPHINE
AMIDE*1	PHOSPHONATE
AMIDE*2	EPOXIDE
AMIDE*3	ETHER
ISOCYANATE	PEROXIDE
ACID*HALIDE	ALCOHOL
THIOESTER	NITRITE
AMINE*3	O*SULFONATE
AZIRIDINE	FLUORIDE
AMINE*2	CHLORIDE
AMINE*1	BROMIDE
NITROSO	IODIDE
DIAZO	DIHALIDE
HALOAMINE	TRIHALIDE
HYDRAZONE	ACETYLENE
OXIME	OLEFIN
IMINE	HYDRATE
THIOCYANATE	HEMIKETAL
ISOCYANIDE	KETAL
NITRILE	HEMIACETAL
AZO	ACETAL
HYDROXYLAMINE	AZIDE
NITRO	DISULFIDE
AMINEOXIDE	ALLENE
THIOL	LACTONE
EPISULFIDE	VINYLW
SULFIDE	VINYLD
SULFOXIDE	ESTERX
	AMIDZ

example, suppose a carbon-nitrogen triple bond has been found the questions would look like

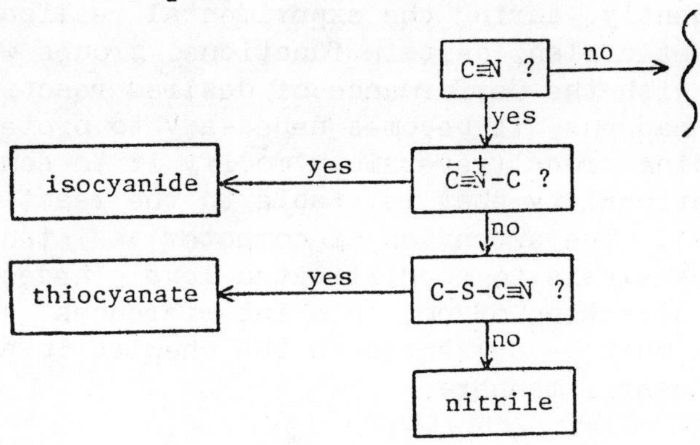

The actual data table which drives this recognition process is reproduced below.

A22	LOC A23	LOC A24	SHIFT + IF CARBON*COUNT IS TWO
A23	NULL	NULL	IDENTIFIED AS KETONE
A24	LOC A25	LOC A26	IF HYDROGEN*COUNT IS TWO
A25	NULL	NULL	IDENTIFIED AS ALDEHYDE
A26	LOC A27	LOC A28	IF HYDROGEN*COUNT IS ONE
A27	LOC A25	LOC A28	IF CARBON*COUNT IS ONE
A28	LOC A29	LOC A32	SEARCH FOR C**N
A29	LOC A30	NULL	SHIFT + SEARCH FOR C*N
A30	LOC A31	LOC A31	NONORIGIN ENTRY
A31	NULL	NULL	IDENTIFIED AS ISOCYANATE
A32	LOC A33	LOC A33	ENTRY BOND*SHARED
A33	LOC A34	LOC A43	SEARCH FOR C*O
A34	LOC A35	LOC A35	BOND*SHARED
A35	LOC A36	LOC A37	SHIFT + IF HYDROGEN*COUNT IS ONE
A36	NULL	NULL	IDENTIFIED AS ACID
A37	LOC A38	NULL	SEARCH FOR C*O
A38	LOC A39	LOC A39	SHIFT + NONORIGIN
A39	LOC A40	LOC A40	BOND*SHARED
A40	LOC A41	LOC A42	SHIFT + IF IN RING OF ANY SIZE
A41	NULL	NULL	IDENTIFIED AS LACTONE
A42	NULL	NULL	IDENTIFIED AS ESTER
A43	LOC A44	LOC A45	SEARCH FOR C*X
A44	NULL	NULL	IDENTIFIED AS ACID*HALIDE
A45	LOC A46	LOC A66	SEARCH FOR C*N
A46	LOC A47	LOC A47	BOND*SHARED
A47	LOC A48	LOC A49	SHIFT + IF HYDROGEN*COUNT IS TWO
A48	NULL	NULL	IDENTIFIED AS AMIDE*1
A49	LOC A50	LOC A57	SHIFT + IF IN RING OF ANY SIZE
A50	LOC A51	NULL	SHIFT + SEARCH FOR C*N
A51	LOC A52	LOC A52	SHIFT + NONORIGIN
A52	LOC A53	LOC A53	SHIFT + BOND*SHARED
A53	LOC A54	LOC A56	SEARCH FOR C*N
A54	LOC A55	LOC A55	SHIFT + NONORIGIN
A55	LOC A56	LOC A56	BOND*SHARED
A56	NULL	NULL	IDENTIFIED AS LACTAM
A57	LOC A58	NULL	IDENTIFIED AS LACTAM
A58	LOC A59	LOC A59	BOND*SHARED
A59	LOC A60	LOC A60	SHIFT + NONORIGIN
A60	LOC A61	LOC A64	SHIFT + SEARCH FOR C*N
A61	LOC A62	LOC A62	BOND*SHARED
A62	LOC A63	LOC A63	SHIFT + NONORIGIN
A63	NULL	NULL	IDENTIFIED AS AMIDE*3
A64	LOC A65	LOC A63	IF HYDROGEN*COUNT IS ONE
A65	NULL	NULL	IDENTIFIED AS AMIDE*2

FUNCTIONAL GROUP REACTIVITY

Frequently, during the experimental realization of a synthetic plan, certain functional groups will interfere with the performance of desired reactions. When this happens, it becomes necessary to protect the offending group (reversibly modify it to some other functionality that is stable to the reaction conditions). The extension of computer assisted synthetic analysis to sophisticated levels necessitates the detection of possible interferences. Such situations must be presented to the chemist in a generally useful manner.

This problem has been attacked in LHASA by the separation of functional group into different classes based on their electronic and steric environment. At the same time a library of standard reagents (currently 60) has been prepared containing the stability of each of the classes of functional groups to each reagent. By this mechanism the program can decide whether groups of an identical or similar type will interfere with the transform. For example, in the structure below it is possible to selectively hydrogenate bond A in the presence of B

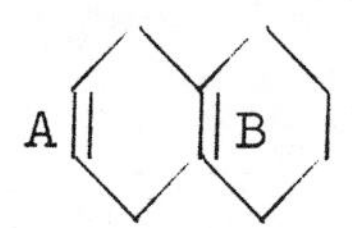

From a computational point of view, functional group reactivity is straightforward. Associated with each group origin is a number which unambiguously defines the environment of the origin. These include steric hindrance and accessibility, strain, and electronic environment. It is important to note that a group can be in several subclasses simultaneously, all of which must be encoded into the one number. This number is used to assign reactivity levels to each origin relative to each reagent.

At the time of attempted reaction execution, the program reads the desired conditions from the data base. The reactivity levels of all non-participating groups are examined. If all are less reactive than the participating group(s), then the transform is allowed to proceed. If this is not the case, the offending group(s) is examined to see if it is generally protectable. If it is then a solid rectangle is drawn around the group as the transform is displayed to the chemist. Unstable, unprotectable groups are graphically indicated by a dashed box around their bonds.

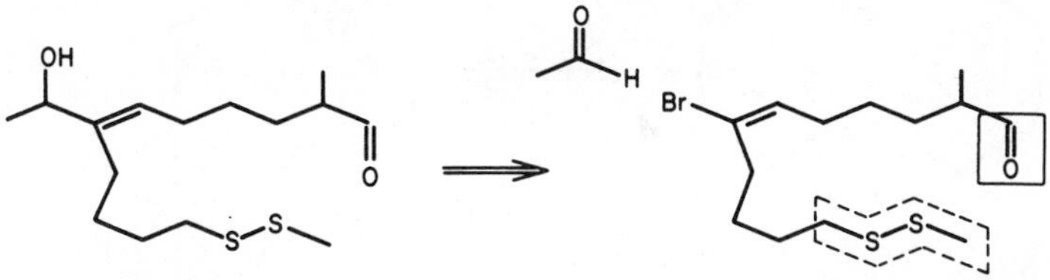

LHASA does not try to assign specific protecting groups. There is just so much chemical detail that would have to be programmed that the interactive aspect of LHASA would be severely degraded. An additional problem entails evaluating when in the synthetic route it would be best to protect and then deprotect the group(s). The program currently deals with the synthesis tree on a node by node basis. A global optimization of the individual steps in the tree is one additional level of sophistication which has not yet been attempted.

APPENDAGE BASED STRATEGIES

The vast majority of multistep syntheses involve either the disconnection, reconnection, or modification of what are loosely called 'appendages'. One particularly useful retrosynthetic strategy consists of fragmenting a ring and then disconnecting the resulting appendages, as shown below.

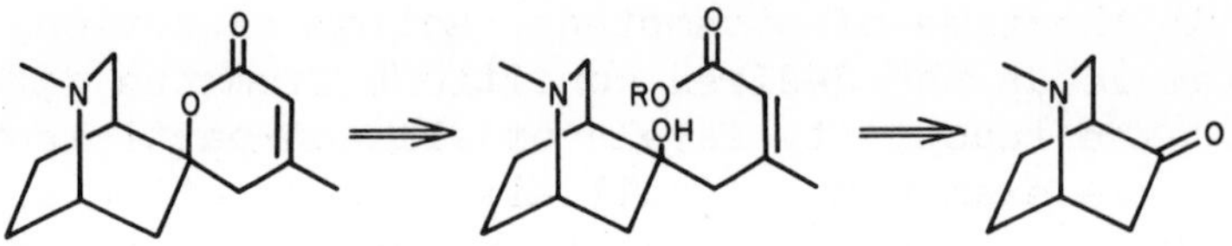

Similarly, strategies involving reconnection of appendages are exceptionally useful in stereospecific synthesis. These reconnections have also proven valuable in the synthesis of medium size rings.

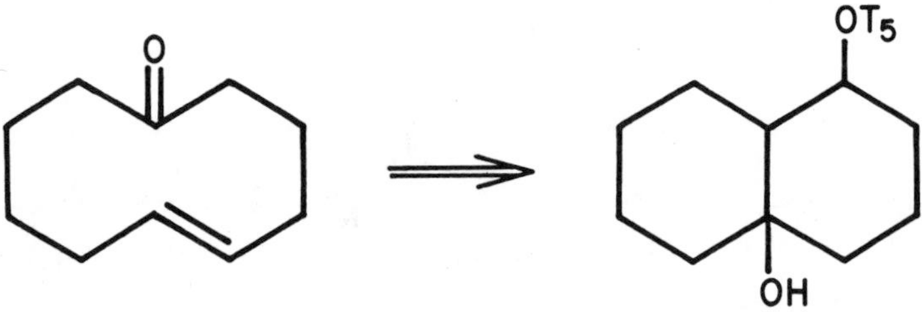

It is important to note that all stereochemistry in these examples is perceived by LHASA and used in its strategies.

There are two classes of appendages - ring appendages and branch appendages. A ring appendage is a group of atoms attached to ring that is not in a ring itself. A branch appendage may only originate on a non-ring atom and must have three or more attachments other than hydrogens. Non-terminal olefins and acetylenes are also considered as origins of branch appendages for chemical reasons. Significant in the use of appendages is the combinatorial problem of determining identicality of appendages. This has been solved quite elegantly by Jorgensen.[4]

Appendage based strategies may be divided into disconnective and reconnective. The latter may be further partitioned into ring appendage - ring

---

[4] Corey, E. J., W. L. Jorgensen, J. Amer. Chem. Soc., 98, 189 (1976).

appendage, ring appendage-ring, and acyclic reconnections. LHASA currently knows about twenty different classes of reconnective transforms - a small subset of the group pair chemistry data base (vide infra). When in a mode where these transforms are specifically being executed, they are empowered to make several small structural modifications to achieve the desired reattachment.

THE CHEMISTRY PACKAGES OF LHASA

Functional Group Based Transforms

As we have already seen, LHASA has a wide variety of strategies which it can employ, either of its own volition or by directive of the chemist-user. In order to facilitate the use of these strategies, the chemical data base in LHASA is broken down into several separate categories, two group transforms, one group transforms, functional group interchange, functional group addition and ring oriented transforms. This section will describe each of these and give brief examples of their use.

Two group transforms are keyed specifically by two functional groups with a path of predetermined length between them. Examples of these are shown below.

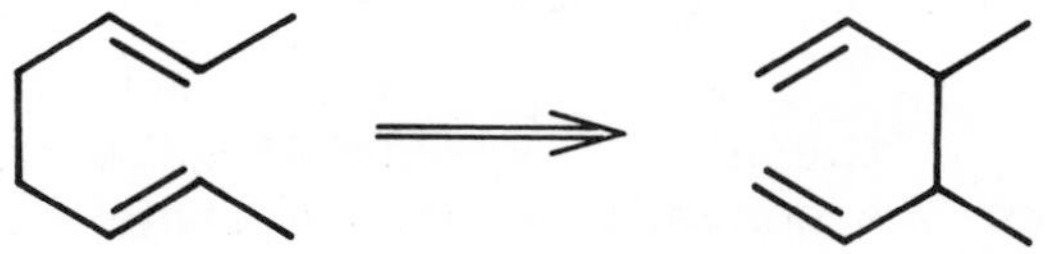

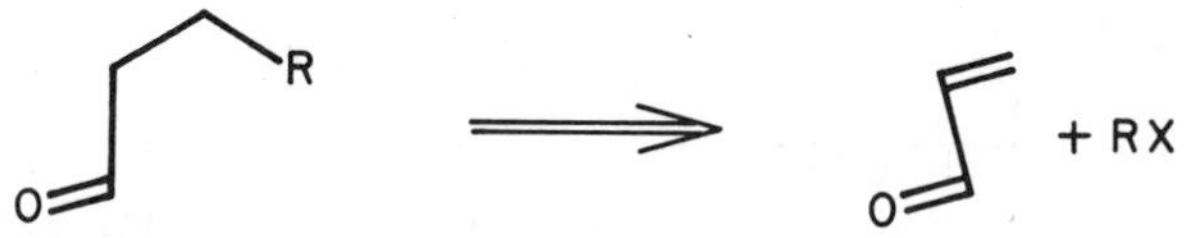

R

O

O + RX

One group transforms are similar but are keyed by one specific group with an associated path (not as chemically meaningful as above). Examples are shown below.

OH R → O + RX

HO →

The world of chemistry would indeed be rosy if there were always precise matches against this data base. Unfortunately, this is not often the case. Frequently, one of the group (in a two group situation) does match, but the other one does not. If the incorrect group could readily be converted into a matching group, then the transform would become acceptable. As in the Aldol Condensation, if the molecular fragment present were the ether, the match would not be found, yet the etherification of the alcohol can often be quite straightforward. If the

RO → HO → O O

performance of the Aldol is considered a chemical 'goal', then the conversion of the ether to the alcohol is a 'subgoal'. In this case, the subgoal consisted of modifying a group or Functional Group Interchange (FGI).

A more complicated case would exist if the second group necessary to key the transform was totally absent. In the example below, the only functional substructure capable of keying a transform would be the olefin. To perform the Aldol transform it would be necessary to add (in the retrosynthetic direction) the alcohol group. Such subgoals are called Functional Group Addition (FGA). Obviously one does not want to always introduce all possible functional groups at all available positions or do indiscriminant group conversions without some guiding purpose or strategy. As such, FGI's and FGA's are only executed in response to a request from a higher level chemistry package.

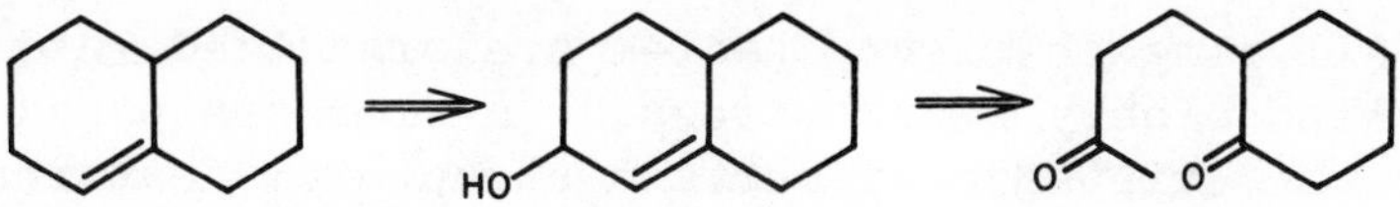

Subgoal requests can be combined and mixed according to the situation. Next to be added is FGI then INTRO since not all groups may be INTRO'ed.

Ring Oriented Transforms

It was recognized that of great significance to LHASA type analyses was the inclusion of chemistry packages whose sole purpose was the construction of rings. These transforms could not and should not be keyed by the presence or absence of any particular functionality. Since they had specific long range goals, they were given considerable power in the type and number of subgoals that they could request. This is in contrast to the two group or one group chemistries where only one FGI or FGA could be performed before the final disconnection.

Four ring forming transforms have been considered at length by the LHASA development group - the Diels Alder addition, the Robinson Annelation, the Simmons-Smith reaction, and iodo-lactonization. The first three of these have been fully implemented in LHASA and the fourth is completely flow charted and awaits only coding into the chemistry data base language.

All ring chemistry tables are organized into what is called binary search trees. Queries are posed about the existence of certain structural features. Each of these questions is answerable with a yes or a no. Based on the answer one of two different follow-up questions is selected. Embedded within the table may be requests for subgoals, either those already in the FGI or FGA table or for special reactions which are needed only for these transforms and are not of general synthetic interest.

The first step in implementation of a ring transform is the preparation of a chemical flow chart. This defines all the questions about the structure and describes in a graphic representation the synthetic steps that will be taken. It is quite straightforward for a chemist having no familiarity with LHASA to read and make use of these charts. A number of graduate students and postdoctoral fellows in the Corey group at Harvard University made significant input to the chemistry in the tables without ever having to worry about the computer implementation.

The example below shows some of the synthetic routes generated by the Diels Alder transform for the indicated precursor. It is important to note that while some of the chemistry may look somewhat naive, it can be quite thought provoking.

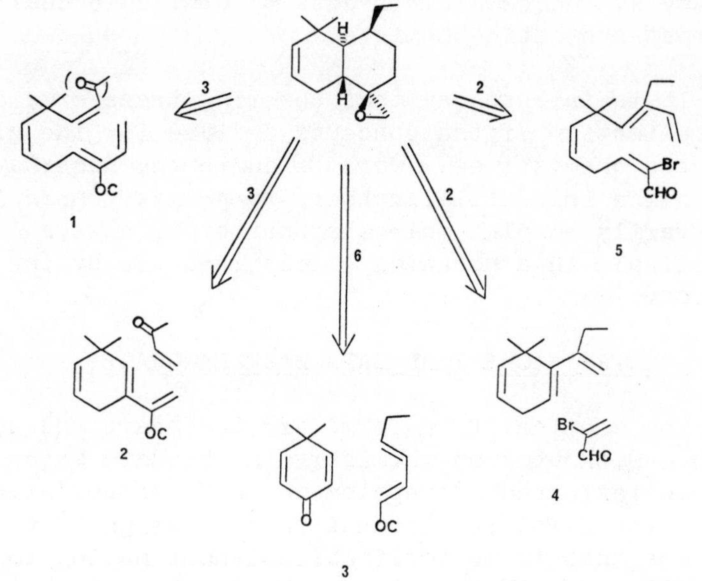

H
H
O
3
OC
1
2
Br
CHO
5
3
6
2
OC
2
O
OC
3
Br
CHO
4

In terms of newer chemistry, the sample sequence shown below was generated by the iodo-lactonization transform.

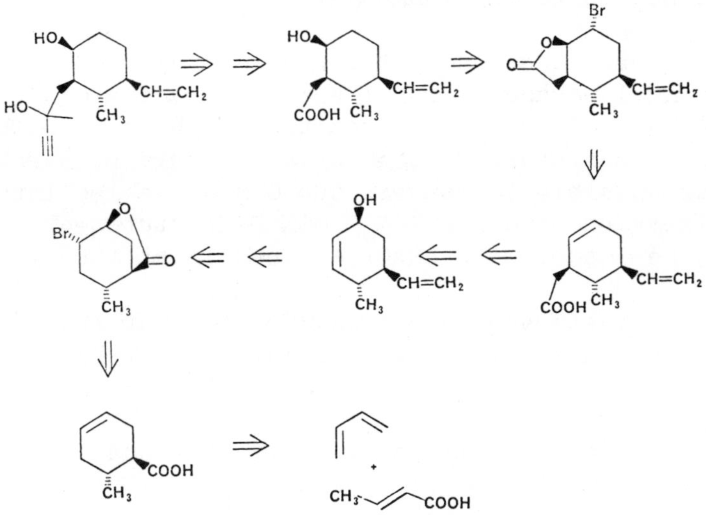

HO
HO
CH=CH₂
CH₃
HO
COOH CH₃
CH=CH₂
Br
O
O=
CH=CH₂
CH₃
Br
O
=O
CH₃
OH
CH=CH₂
CH₃
CH=CH₂
COOH CH₃
COOH
CH₃
CH₃
COOH

It is clear that designing a synthesis of a ring with so many stereocenters presents a formidable challenge for most synthetic chemists.

It is fair to say that the ring transforms are a generalization of the concepts derived for the group oriented chemistries. Work is currently underway to generalize this still further, to permit generation of arbitrarily complex molecular patterns, always specifiable in a notation easily readable by the chemist.

CHMTRN - CHEMICAL DATA BASE LANGUAGE

The chemical transforms are the heart and soul of LHASA. Without good chemistry in the data base, all the sophisticated perception would be essentially useless. The first requirement in the design of the data base was that it be modifiable without having to recompile any other part of the program. The second requirement was that it require no knowledge of FORTRAN or how LHASA is organized on a subroutine by subroutine basis. The third requirement was that the data base be easily readable by chemists with no training in LHASA and modifiable after only a little introduction to the language.

To meet these conditions a special chemical programming language CHMTRN (Chemical Translator)[5] was developed. By use of a special assembler - TBLTRN (Table Translator, written by Dr. Donald E. Barth), it was possible to convert the CHMTRN tables into specially encoded FORTRAN BLOCK DATA statements which could be loaded with LHASA or read in at run time.

The basic approach of CHMTRN is that there are keywords (currently several hundred) that have

---

[5] If this name conflicts or duplicates that of some other chemical program, I apologize. The duplication is unintentional.

specific numerical values assigned to them. All of these keywords that are typed on a single line in the data base are logically 'or' ed together into one computer work. (The translation and combination are handled by TBLTRN). Each such line is called a qualifier as it limits or modifies the scope of the transform.

LHASA contains an interpreter called EVLTRN (Evaluate Transform) which decodes the bit patterns and performs the requested queries about the current structure or performs a specified operation. As an example, consider a line from the tables which says

SUBTRACT 20 FOR EACH PRIMARY HALIDE ALPHA TO CARBON*1 OFFPATH ONRING.

This qualifier decrements the base rating of the transform for each primary halide that is on a cyclic carbon which is not a part of the path keying the transform. From this example, it is possible to see how densely the chemical data is packed (one qualifier takes up only one computer word - 32 bits). There is a target to be searched for (the halide), a domain or location to which the search is restricted (alpha to the carbon but on a ring and off the path), and an iteration command indicating that the operation (the subtraction) is to be performed for each occurrence.

CHMTRN has several other constructions worthy of mention. The first is the ability to make modifications to the structure according to results of qualifier evaluations. One can say, for example,

ATTACH AN ALCOHOL TO CARBON*2 CIS TO CARBON*4

This command also shows just one case where stereochemical considerations can be included.

Complete block structuring (as in PL/I or ALGOL) has been incorporated. This is useful where a complex series of queries should be applied in certain repeatable circumstances, for example,

```
FOR EACH KETONE ANYWHERE DO
    BEGIN
    .
    .
    .
    END.
```

All qualifiers between the BEGIN and END are executed for each KETONE. These may be nested to any desired depth. Similarly IF-THEN-ELSE constructions are also allowed.

Software subroutines have also been implemented in CHMTRN/EVLTRN. Suppose there is one group of qualifiers which needs often be applied to different locations in the molecule at varying times. This can be handled by the construction

```
CALL FGIW AT CARBON*3 AND BOND*2
```

In the subroutine FGIW these arguments are addressable as SPECIFIED*ATOM and SPECIFIED*BOND. It is permissible to apply stereochemical constraints to arguments at the time of execution of the CALL. There is no practical limit on the depth of subroutine calls. Subroutines may also return a value to indicate whether or not they succeeded in the task they were assigned to do.

The Robinson Annelation transform has received detailed examination by the LHASA group. One subroutine in the table specifically checks to see if there is any functionality alpha to the ketone in the cyclohexane and if there is, remove it by exchanging it for something non-offensive. This subroutine is reproduced below as an example of the CHMTRN language.

```
...THIS SUBROUTINE IS CALLED TO CLEAR AWAY ANY UNDESIRABLE FUNCTIONALITY
...ALPHA TO A KETONE ON THE RING

ALPHCHK IF NO HYDROGEN ON THE SPECIFIED ATOM THEN GO TO I9
        IF THERE IS NOT A WITHDRAWING GROUP ON THE SPECIFIED ATOM THEN GO TO I8
        IF THE SPECIFIED ATOM IS THE SAME AS CARBON*2 THEN RETURN SUCCESS
        IF BOND*5 IS A FUSION*BOND THEN RETURN SUCCESS
        IF THERE IS NOT A NITRO ON THE SPECIFIED ATOM THEN GO TO I8
        EXCHANGE THE GROUP FOR AN AMINE
        IF SUCCESSFUL THEN GO TO J2 OTHERWISE RETURN FAIL
I8      IF THERE IS A HALIDE ON THE SPECIFIED ATOM THEN GO TO J2
        IF THERE IS A KETONE ON THE SPECIFIED ATOM THEN GO TO J2
        IF THERE IS A WITHDRAWING GROUP ON THE SPECIFIED ATOM THEN GO TO J2
        IF THERE IS A DONATING GROUP ON THE SPECIFIED ATOM THEN GO TO J2
        IF THERE IS AN OLEFIN ALPHA TO THE SPECIFIED ATOM THEN GO TO J2
        IF THERE IS A FUNCTIONAL GROUP ON THE SPECIFIED ATOM THEN RETURN FAIL
        IF THERE IS NOT A FUNCTIONAL GROUP ALPHA TO THE SPECIFIED ATOM THEN RETURN SUCCESS
        EXCHANGE THE GROUP FOR A WITHDRAWING GROUP
        IF SUCCESSFUL THEN GO TO J2 OTHERWISE RETURN FAIL

I9      IF THE SPECIFIED ATOM IS A QUATERNARY*CENTER THEN RETURN FAIL
        IF THERE IS AN ALCOHOL ON THE SPECIFIED ATOM THEN GO TO J2
        IF THERE IS A HALIDE ON THE SPECIFIED ATOM THEN GO TO J2
        IF THERE IS NOT AN ETHER ON THE SPECIFIED ATOM THEN RETURN FAIL
J2      CALL DET THE GROUP AT THE SPECIFIED ATOM AND GO TO RET
```

Two examples of how these constructs are applied together will demonstrate their utility and flexibility. A number of reactions, such as Michael addition depend on the conformation of the intermediate enolate for their specificity. It is possible to make initial queries about the structure, generate the enolate, ask about it, then generate the final precursor and ask questions about it. At each stage of this process, it is possible to detect a fatal condition and terminate evaluation of the transform.

This same language is being used to successfully calculate preferred conformations of cyclohexanes for evaluation of regiospecificity and in functional group reactivity analysis.

ORGANIZATION OF THE DATA BASE

The one group and two group and the subgoal tables are queried very frequently during a typical analysis session. A data structure has been developed which is extremely efficient for these searches - the retrieval time being independent of either the size of the data table or the number of successful hits in the table. Because of the general applicability of this technique, we shall describe it in more detail - using the two group tables as an example.

At present there are sixty-four different functional groups which are capable of keying a transform in some manner. This can either be as the first keying group or the second group (for example, both a ketone and an olefin key the Aldol Condensation)

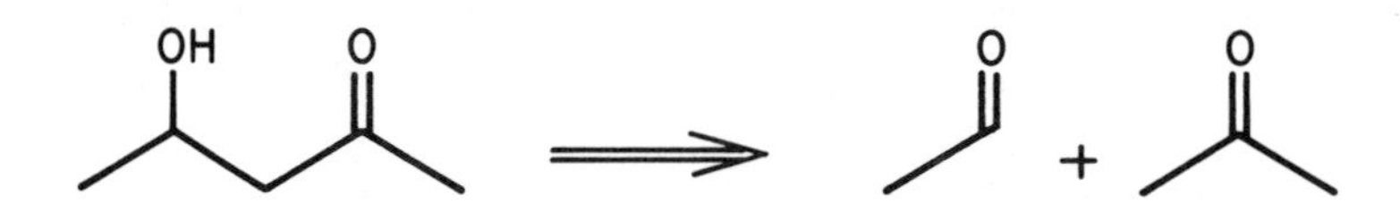

The keying mechanism must point to the applicable transform regardless of the ordering of the groups and it must also handle situations where the keying groups are the same. The first element in the representation is a 'directory set' - a Boolean set with a bit on for each group that can participate in any transform at the desired path length. If the group is not marked in this set, then there is no need to further interrogate the table - there will not be any acceptable entries.

For each group that does participate in transforms, there are two additional multi-word sets. A bit is on for each transform in which the group takes part - in the first set if it is the first keying group and in the second if it is the second. Logically 'AND'ing the first group's first set with the second group's second set yields a set with bits on for only these transforms. These bit positions are used as indexes into a table of addresses of the qualifiers for those applicable transforms. While it sounds rather complicated, it really is not. What has been done is to generate an external addressing structure at assembly time.

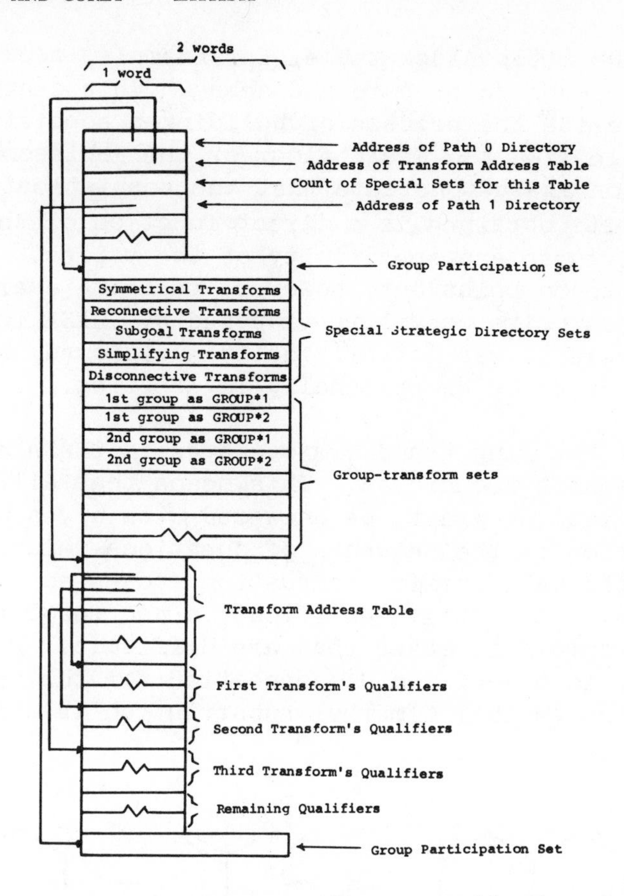

The figure above shows the overall structure of the table. It should be noted that if you wish to restrict your search to, for example, those transforms which break carbon-carbon bonds all that is necessary is to define, at assembly time, a set to indicate this characteristic and indicate which transforms are applicable. At run time, 'AND'ing this set with otherwise allowed transforms applies the restriction in parallel. This technique of generating an external addressing structure when coupled with Boolean operations is a quite powerful and useful technique.

HOW DO YOU ADD TO THE DATA BASE

The criticism has often been levelled at LHASA that it takes considerable time to add to the data base. The Diels Alder table, for example, took almost six man-months to prepare and debug. This section will describe the process of building a sophisticated data table like the Diels Alder or the Robinson Annelation and demonstrate that the sophistication of the results obtained is a direct function of the exhaustiveness and specificity of the tables. (It is worthwhile to point out, however, that there are often times when naive chemistry proposed by LHASA in situations where it was not originally envisioned, has turned out to be exceptionally interesting.)

All the ring transform packages in LHASA employ binary search techniques. This means that all structural questions are to be answered with a yes or a no. Preparation of the sequence of questions relating to straightforward chemical situations poses no real problems. It is the identification and resolution of the extraordinary cases that are difficult. For example, in a Robinson disconnection for the sequence below, the geminal dimethyl substitution is a formidable problem.

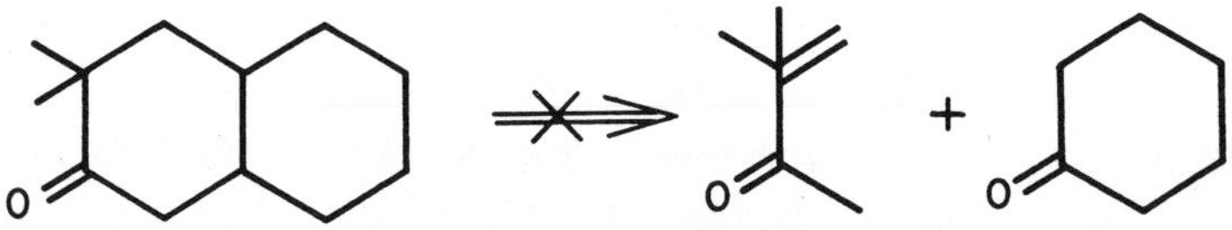

It is up to the chemist designing the tables to first perceive that this situation might occur. Second, decide whether he wishes to have the tables salvage the difficult situation and if he does, he has to manually determine what kind of chemistry should be attempted.

The above example is a clear black or white situation. Unless the dimethyl substituent is removed, the transform just cannot proceed. The grey areas cause just as much of a dilemma for the chemist. In Marshall's synthesis of isonootkatone two possible stereoisomers could have resulted.

COOMe O O ⟹ ⟹ O

A priori prediction of the stereochemical course of a reaction, even knowing the three dimensional structure of the reagents is quite difficult, if not impossible. When preparing a section of the table dealing with such an ambiguity the chemist is faced with two alternatives, disregard stereochemistry entirely (and make sure that LHASA does not imply that any stereochemistry is being specified) or go into the laboratory and run an experiment. The recognition of this situation can often get the table writing chemist thinking and has sometimes even suggested specific reactions that should be run.

As we have been adding to the data base at Du Pont (to the one and two group tables), the question has often been raised "how much detail should we go into in the qualifiers?" This is somewhat of a dilemma, many of the industrial reactions we are dealing with have only been considered for a limited number of substrates. It is not clear whether qualifiers should be incorporated that restrict the transforms to only those cases where it is known to work or whether only those which are known to fail should be specifically excluded. Both ways take an immense amount of literature work to do consistently.

We all know that butadiene can be dimerized under catalytic conditions to a wealth of different products. Addition of alkyl substituents changes the product mix and introduces a variety of different stereoisomers as well. What happens if we put functional groups on butadiene. Do all the reactions still proceed - do you get any new ones, etc.? We do not know and serious doubt whether many experiments have ever been run on such systems. The second alternative above has been chosen - qualifiers are being used only to exclude reactions known not to work. As such, a lot of naive chemistry comes out of our version LHASA, some of it exceptionally poor but at least the analyses are reasonably comprehensive.

In summary, adding simple reactions to LHASA is simple. Incorporating sophisticated reactions can be as complicated as you wish to make it. (Work is currently underway to prepare a general package of subgoal transforms which will serve to remove interferences - relieving the chemist from having to work them out separately for each super transform.)

CONCLUSION

Why is Du Pont interested in LHASA? The program was clearly designed for carbocyclic natural products synthesis in mind - an area in which the Company has only limited interest.

- Our attempt to add industrial synthetic chemistry to LHASA is forcing us to organize our thoughts along lines heretofore not done. We are being required to look at our reactions in terms of their known <u>and</u> <u>unknown</u> generality. This in and of itself is highly beneficial.

- Our pharmaceutical and agrichemical chemists have been using the natural products aspects of LHASA to generate new ideas, often not of industrial synthetic merit but certainly of interest when looking for ways to make derivatives, especially commonality of routes.

- Lastly, we have already seen that some of our industrial knowledge is turning out to be useful to organic synthesis in other areas. For example, very few chemists outside of those who are actually using it daily are aware that, given suitable catalysts, butadiene can dimerize into the compound below - a quite attractive (and inexpensive) starting material for prostaglandin synthesis.

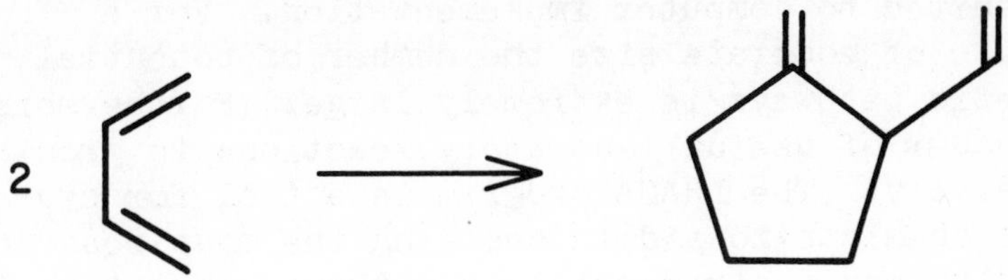

Our hope is that LHASA will help us to insure that we have considered all reasonable routes to our major products.

ACKNOWLEDGMENTS

The LHASA project has been in existence since 1968. During the years a number of extraordinately able graduate students and post doctoral fellows came to work with Professor Corey, almost all laboratory synthetic chemists. Whether it is the infectious enthusiasm for the project or just an enjoyment of using the computer as a research tool, not one alumnus of the LHASA project has abandoned his involvement with computers and returned to bench chemistry. They are (with current locations)

W. J. Howe - Upjohn
D. E. Barth - Harvard Business School
W. L. Jorgensen - Purdue
R. D. Cramer - Smith Kline and French
W. Todd Wipke - Santa Cruz
G. A. Petersson - Wesleyan
J. W. Vinson - Harvard
H. W. Orf - Harvard

ABSTRACT

Design of complex organic syntheses is a task well suited to computer implementation. For a molecule of moderate size the number of potential synthetic pathways is extremely large. Furthermore the number of useful laboratory reactions is growing explosively. The LHASA program is a tool for synthetic chemists to aid in choosing the most reasonable routes to any desired molecule without exhaustive enumeration. The basic structure of the program and the chemistry it employs are discussed giving special consideration to the strategies employed in selection of routes.

2

Computer Programs for the Deductive Solution of Chemical Problems on the Basis of a Mathematical Model of Chemistry

JOSEF BRANDT, JOSEF FRIEDRICH, JOHANN GASTEIGER, CLEMENS JOCHUM, WOLFGANG SCHUBERT, and IVAR UGI

Organisch-Chemisches Institut, Technische Universität München, Arcisstr. 21, D-8 München-2, Germany

1. Introduction

1.1 Computers and Progress in Chemistry

In the past decades there has been dramatic progress in chemistry. The conceptual and theoretical approaches to chemistry have been decisively shaped by quantum mechanics. The research topics and techniques of chemistry have been changed in a profound manner by the advent of modern separation and purification methods as well as advances in structural determination by the various types of spectroscopy and X-ray crystallography. The use of computers in chemistry has also played an essential role in this context. However, it is safe to predict that the major contributions of computers to chemistry are still to be expected in the future.

Computers are already an important tool of chemistry. Computer-assisted documentation, the collection and evaluation of experimental data, and quantum mechanical calculations of molecular properties predominate. The importance of the solution of chemical problems such as the design of syntheses with the aid of computers is not yet widely recognized, but will have far-reaching consequences and may well lead to rather fundamental changes in the activity of chemists. In particular, the computer programs for the deductive solution of chemical problems on the basis of logical structure models and mathematical representation of chemistry have great potential for the future.

1.2 The Solution of Chemical Problems on the Basis of Theoretical Physics and Mathematics

The molecular systems, the subject of chemistry, consist of a rather small number of building blocks, namely the ca. 100 chemical elements which are combined according to rules derived from mathematically stated physical principles. The representation of chemistry in mathematical terms seems therefore quite natural. This has not yet been utilized to the full conceivable extent.

The energy hypersurface describes the energy of chemical systems with a given set of atoms as a function of the spatial distribution of the atomic nuclei. If one could compute the complete energy hypersurface for any set of atoms and analyse the corresponding data in a suitable manner, one could predict most of the relevant properties of molecular systems with the methods of theoretical physics and mathematics.

This, however, is not feasible in the foreseeable future, because the computational effort would be extremely large, even in the case of rather small molecular systems.

For the solution of many chemical problems the theoretical treatment of a few selected points and their vicinity on the energy hypersurface would suffice. Yet, in many cases, one does not know which points are the relevant ones. For chemistry it would be rather useful to have a method for providing a survey of those points and pathways on an energy hypersurface which are essential for the solution of a given problem, without the necessity of a quantum mechanical treatment of wide areas of an energy hypersurface.

In the present paper a simple mathematical model of the chemistry of a given set of atoms is presented which affords precisely the latter, and may serve as a basis of computer programs for the deductive solution of a great variety of chemical problems.

2. The Chemistry of a Fixed Set of Atoms

2.1 Chemical Equivalence Classes

Any progress in chemistry may be interpreted as the recognition of new equivalence relations and classes. Until recently, a universal model of chemistry could not be defined, because we lacked the concepts for a mathematical treatment of the constitutional aspect of chemistry. Therefore it is necessary to consider any constitutionally different molecular systems as equivalent regardless of the number of molecules which they contain, if, in principle they are interconvertible through chemical reactions.

2.2 Isomeric Ensembles of Molecules

The set of all molecules can be partitioned into equivalence classes whose members have all the same molecular formula, i.e. the equivalence classes of isomers. Isomerism is here the relevant equivalence relation.

Extending the concept of isomerism from molecules to ensembles of molecules (EM) leads to a mathematical model of constitutional chemistry[1]. An EM is described by its list of the molecular species in the form of suitable chemical formulas. For an EM one can define two types of empirical elementary formulas, on one hand the ensemble formula, on the other hand a partitioned empirical ensemble formula which consists of the molecular formulas of the molecules in the EM.

Let $\underline{A}$ be a set of atoms. Then all EM($\underline{A}$), i.e. all EM which can be made from $\underline{A}$, have the same ensemble formula $\langle\underline{A}\rangle$. The partitioned empirical formula of an EM is a partitioning $\{\langle A_1\rangle,\ldots\langle A_s\rangle\}$ of $\underline{A}$ in which each $\langle A_i\rangle$ is the formula of a molecule. Accordingly, an EM($\underline{A}$) consists of one or more molecules, which are obtained from $\underline{A}$, if each atom of $\underline{A}$ is used exactly once.

Since isomeric molecules may differ constitutionally and stereochemically, a partitioned ensemble formula generally corresponds to more than one EM($\underline{A}$). The constitutional formula of an EM($\underline{A}$) contains the constitutional formulas of all molecules in that EM($\underline{A}$). An FIEM($\underline{A}$), the family of isomeric ensembles of molecules of an atom set $\underline{A}$ is the collection of all

EM(A). An FIEM is given by the empirical formula of the underlying set of atoms A.

A chemical reaction or a sequence of chemical reactions, is the transformation of an FM into an isomeric EM. Thus the whole chemistry of a set of atoms A is given by the EM(A) and their interconversions within the FIEM(A).

An energy hypersurface describes all chemical systems which contain a given set of atoms. An FIEM(A) of stable EM(A) corresponds to a family of energy minima on the energy hypersurface of the atom set A. Sequences of EM(A) which are chemically interconverted correspond to pathways on the energy hypersurface of A.

3. BE-Matrices

In molecular systems the atomic cores are held together by valence electrons which occupy molecular orbitals involving two, or more atoms.

A covalent bond is a pair of valence electrons in a molecular orbital about two cores. Generally, the chemical constitution of a molecular system is described by its covalently bound pairs of atomic cores. The distribution of those valence electrons which are not contained in covalent chemical bonds can be included in the description of a chemical constitution.

A chemical constitution is usually represented by a constitutional formula in which the chemical element symbols correspond to the respective atomic cores, the covalent bonds are indicated by lines between the element symbols, and the free valence electrons by dots at the atomic symbols.

The chemical constitution of molecules has also been represented by various types of matrices. For the present purpose the BE-matrices are particularly suitable.

An n x n BE-matrix

$$B = \begin{pmatrix} b_{11} & b_{12} \cdots\cdots\cdots\cdots & b_{1n} \\ b_{21} & & \\ & & \\ n_{n1} & \cdots\cdots\cdots\cdots\cdots & b_{nn} \end{pmatrix} \qquad (13)$$

refers to a molecule or EM which contains an atom set $\underline{A} = \{A_1,...,A_n\}$ with n indexed atoms. The i-th row and i-th column of a BE-Matrix belongs to the i-th atom of $\underline{A}$.

An off-diagonal entry b_{ij} in the i-th row and j-th column is the formal covalent bond order between the atoms A_i and A_j. Since this implies also a bond from A_j to A_i, we have $b_{ij} = b_{ji}$. Thus BE-matrices are symmetric.

The i-th diagonal entry of a BE-matrix is the number of free valence electrons which belong to A_i.

The number of valence electrons which belong to the atom A_i is given by the row-column sum

$$b_i = \sum_j b_{ij} = \sum_j b_{ji}$$

The cross sum $\hat{b}_i$ over a diagonal entry comprizes all entries in the i-th row and columns and is equal to

$$\hat{b}_i = 2b_i - b_{ii}.$$

The cross sum $\hat{b}_i$ is the number of valence electrons which occupy the valence orbitals of A_i.

A table which contains for all chemical elements the allowable combinations of b_i, $\hat{b}_i$, coordination numbers and bond orders affords a quick check whether a BE-matrix represents a valence chemically stable molecular system.

The sum

$$S = \sum_{ij} b_{ij}$$

over all entries of a BE-matrix is the total number of valence electrons in the EM. For all EM of an FIEM this number is the same.

Unless the indexing of the atoms is fixed according to some rule, a given chemical constitution is not only represented by one BE-matrix, but there are up to n! equivalent BE-matrices which differ by permutations of the atomic indices and the respective row/column indices.

Any BE-matrices B and B' which differ only by row/column permutations describe the same EM. The transformation of an n x n BE-matrix B by row/column permutation into an equivalent BE-matrix B' can be achieved by an n x n permutation matrix P and its inverse, P^t, according to

$$B' = P^t \cdot B \cdot P$$

For various purposes, such as documentation, one needs an unambiguous correspondence between an EM and its BE-matrix. Therefore, we have developed procedures for the assignment of atomic indices which lead to an canonical BE-matrix.

Among the equivalent BE-matrix of an EM which consists of several molecules $M_1, \ldots, M_m$, there exist BE-matrices in block form, such that each block represents one contiguous molecule. The block form of BE-matrices is useful for the separation of molecules of EM whose BE-matrices have been generated by a computer.

4. The Representation of Electron Relocation by R-Matrices

A chemical reaction is the conversion of an EM into an isomeric EM by relocation of valence electrons.

Since the total number of valence electrons S does not change in a chemical reaction, it must be the same for the initial EM(B) and the final EM(E) of a chemical reaction. It follows that the sum of entries of BE-matrices must be invariant under those BE-matrix transformations B → E which represent chemical reactions EM(B) → EM(E).

Let B and E be the BE-matrix of the starting EM(B) and the final EM(E) of a chemical reaction. Then an R-matrix (reaction matrix) is defined by the transformation

$$B + R = E.$$

The sum of the entries r_{ij} of an R-matrix is

$$\sum_{ij} r_{ij} = 0$$

because

$$s \sum_{ij} e_{ij} = \sum_{ij} b_{ij} + \sum_{ij} r_{ij} = \sum_{ij} b_{ij}$$

The matrix R = E - B must be symmetric because the BE-matrices B and E are symmetric by definition.

The matrix $\bar{R}$ = -R is the inverse of R. The transformation E + $\bar{R}$ = B represents the reaction EM(E) → EM(B), i.e. the retro-reaction of EM(B) → EM(E).

An R-transformation B + R = E represents a chemical reaction if, and only if it obeys the mathematical fitting condition

$$e_{ij} = b_{ij} + r_{ij} \geqslant 0$$

for all entries because, by definition, a BE-matrix must have non-negative entries. Furthermore, the result E of an R-transformation of a BE-matrix B of a stable EM(B) will represent a stable EM(E), if the entries e_{ij} of E have values that are permissible for the respective atoms A_i and A_j("chemical fitting").

Accordingly, the positive entries b_{ij} of a given BE-matrix B may be used to determine the negative entries of mathematically fitting R-matrices. The positive entries must be selected to yield an R-matrix with $\sum_{ij} r_{ij} = 0$. Moreover, the entries of an R-matrix can be selected to meet the valence chemical restrictions of the chemical elements in E = B + R.

Thus it is possible to find for a given EM all of its chemical reactions and their products by the transformation properties of the BE-matrices, and the valence chemical constraints of the latter. The application of all mathematically and chemically fitting R-matrices to a BE-matrix generates the BE-matrices of the whole FIEM.

4.2 R-Categories

The transformation of a BE-matrix by a fitting R-matrix corresponds to a chemical reaction.

The off-diagonal negative entries $r_{ij} = r_{ji}$ indicate the number of broken covalent bonds between A_i and A_j, and a negative diagonal entry r_{ii} tells how many free electrons the atoms A_i lose. The positive off-diagonal entries $r_{ij} = r_{ji}$ are the numbers of the newly made covalent bonds A_i-A_j, and the positive diagonal entries r_{ii} correspond to the increases in free electron numbers at the atoms A_i.

Generally, an R-matrix does not only fit one, but many BE-matrices. Accordingly, an R-matrix does not only represent an individual chemical reaction, but a whole category of reactions which have in common the electron relocation pattern represented by the R-matrix[2]. An R-category is an equivalence class of chemical reactions which have in common the same electron relocation pattern and certain features of the participating bond systems. The row/column permutation equivalence of BE-matrices implies that R-matrices represent the same reactions when they are interconverted by row/column permutation according to $P^tEP = P^t(B + R)\,P = P^tBP + P^tRP$.

R-matrices belong to the same R-category if they are interconverted by reduction or expansion, i.e. by removal or attachment of rows and columns containing only zeros.

The elimination of all rows and columns containing only zeros from an R-matrix yields the corresponding <u>irreducible R-matrix</u>.

Any two R-matrices R and R' belong to the same R-category if there exists an R-matrix R'' which can be transformed into R according to

$$R = P^tR''P,$$

and from which R' can be obtained by removal or attachment of rows and columns containing zeros only.

Any two chemical reactions which are represented by the same irreducible R-matrix belong to the same R-category.

With few exceptions, the synthetically important organic chemical reactions proceed with a relocation of electrons involving up to six atoms. In such reactions up to three bonds are broken, and/or newly made, in some cases accompanied by a simultaneous change of the formal electrical charge by +1 at one atom and -1 at another one. Such reactions belong to R-categories whose R-matrices have up to three off-diagonal pairs of positive and negative entries $r_{ij} = r_{ji} = +1$, and $r_{ij} = r_{ji} = -1$. The non-zero diagonal entries $r_{ii} = \pm 2$, corresponding to non-radical reactions, are placed in such a manner that all row/column sums are zero, except one row/column pair with $r_i = \sum_j r_{ij} = \pm 1$.

Table 1 shows such R-matrices as lists of their non-zero entries.
The following reactions (→) or their retro-reactions (⇐) respectively, are examples for the R-categories listed in Table 1.

R-Cat. 1: $F^i\!-\!H^j \longrightarrow F:^{i\ominus} + H^{j\oplus}$

R-Cat. 2: $Li^i\!-\!C^jH_3 + H^{k\oplus} \longrightarrow Li^{i\oplus} + H_3C^jH^k$

R-Cat. 3: $J^i\!-\!C^jH_3 + :O^k\!-\!H^{\ominus} \longrightarrow :J^{i\ominus} + H_3C^j\!-\!O^kH$

R-Cat. 4: $C_6H_5N{=}C\!:^i + Cl^j\!-\!Cl^k \Longleftarrow C_6H_5{=}C^i(Cl^j)(Cl^k)$

R-Cat. 5: $H_3C\!-\!C^k({=}O)\!-\!C^j(\!-\!Cl^i){=}N\!-\!C_6H_5 + :O^l\!-\!H^{\ominus} \longrightarrow H_3C\!-\!C^k({=}O)\!-\!O^l\!-\!H + H_5C_6\!-\!N{=}C\!:^j + Cl^{i\ominus}$

R-Cat. 6: $HC^i{=}C^jH$, $H_2C^k\!-\!C^lH_2$ (four-membered ring, $C^i\!-\!C^k$, $C^j\!-\!C^l$) $\longrightarrow H_2C^k{=}C^iH\!-\!C^jH{=}C^lH_2$

Table 1

The R-Matrices of Some Typical Non-Radical Reactions

R-Category	No. of Broken Bonds	No. of Bonds Made	No. of Participating Free Electrons	Off-Diagonal Entries (with Indices)															Diagonal Entries (with Indices)					
No.				ij	ik	jk	il	jl	kl	im	jm	km	lm	in	jn	kn	ln	mn	ii	jj	kk	ll	mm	nn
1	1	0	1	-1															+2					
2	1	1	0	-1	+1																			
3	1	1	2	-1	+1														+2		-2			
4	2	1	1	-1	-1	+1													+2					
5	2	1	3	-1		-1			+1										+2	+2	-2			
6	2	2	0	-1	+1			+1	-1															
7	2	2	2	-1		+1			-1				+1						+2				+2	
8	3	1	2	-1		-1	-1		+1										+2	+2				
9	3	2	1	-1		+1			-1	-1			+1						+2					
10	3	2	1	-1		+1			-1				+1					-1	+2					
11	3	2	3	-1				+1	-1			-1						+1	+2		+2			-2
12	3	3	0	-1		+1			-1				+1	+1				-1						

and $C^iH_2{=}C^jH_2 + Br^k{-}Br^l \longrightarrow H_2C^iBr^k{-}H_2C^jBr^l$

R-Cat. 7: $HC^k{=}N^l{-}C_6H_5,\ N^j{=}\overset{\oplus}{N}{}^i{=}\overset{\ominus}{N}{}^m \longrightarrow HC^k{-}N^l{-}C_6H_5$ (ring with N^j, N^i, N^m)

and (o-chloronitrobenzene: $C^k{-}N^j{\oplus}(O^{\ominus})(O^i)$, $C^l{-}Cl$) $+ :O^m{-}H \longrightarrow$ ($C^k{=}N^j{\oplus}(O^{\ominus})(O^i{:}^{\ominus})$, $C^l(Cl)(O^m{-}H)$)

and $Cl^i{-}C^jH_2{-}C^kH_2{-}C^lH_2{-}\ddot{N}^m(CH_3)_2 \longrightarrow Cl{:}^{i\ominus} + C^jH_2{=}C^kH_2 + C^lH_2{=}N^m(CH_3)_2$

R-Cat. 9: $H_9C_4{-}N{=}C^i{:}\ \ :C^j{=}N{-}C_4H_9 + O^l{=}C^k(CH_3)_2 \Longleftarrow H_9C_4{-}N{=}C^i{-}C^j{=}N{-}C_4H_9$ (ring with $O^l{-}C^k(CH_3)_2$)

and $H_3C{-}C^l({=}O^m){-}C^k({=}O^j){-}CH_3 + :P^i(OCH_3)_3 \longrightarrow H_3C{-}C^l{=}C^k{-}CH_3$ (ring with O^m, $P^i(OCH_3)_3$, O^j)

R-Cat. 10: $(H_3C)_2\overset{\oplus}{N}{}^i{=}C^j{-}H\ \ H_2C^m{-}H^n$, $H{-}C^k{=}C^l{-}H \longrightarrow (CH_3)_2\ddot{N}^i{-}C^jH\ \ H_2C^m$, $H{-}C^k{-}C^l{-}H + \overset{\oplus}{H}{}^n$

R-Cat. 11: $H_5C_6{-}C^m({=}O){-}C^k({=}N^j{-}O^i{-}SO_2C_6H_5){-}C^l_6H_5 + :O^n{-}H^{\ominus} \longrightarrow H_5C_6{-}C^m({=}O){-}O^n{-}H + :C^k{=}N^j{-}C^l_6H_5 + :\overset{\ominus}{O}{}^i{-}SO_2C_6H_5$

R-Cat.12: $HC^{m}=C^{n}H_2$ (with $HC^{L}=^{K}CH_2$) + $CH_2^{i}=^{j}CH_2$ → cyclohexene ring (HC, $C^{n}H_2$, CH_2^{i}, CH_2^{j}, $C^{K}H_2$, HC^{L})

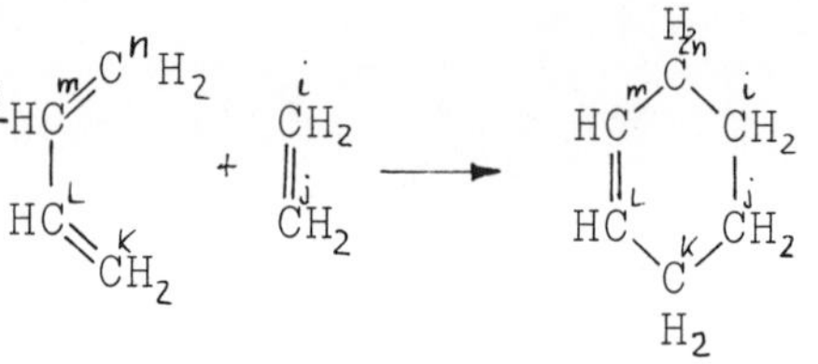

4.3 Reaction Types

Customarily the term "reaction" does not refer to an individual chemical reaction, but a collection of reactions whose members have in common a certain pattern of electron flow involving certain kinds of atoms and bonds, differing only by those moieties of the reacting molecules which do not participate in the process in a direct manner. Thus the term β-elimination comprises a large number of individual reactions, such as

$$CH_3\text{-}CH(OH)\text{-}CH_3 \rightarrow CH_3\text{-}CH{=}CH_2 + H_2O$$

$$CH_3\text{-}CH_2Cl \rightarrow CH_2{=}CH_2 + HCl$$

$$CH_3\text{-}CH{=}CCl\text{-}CH_3 \rightarrow CH_3\text{-}C{\equiv}C\text{-}CH_3 + HCl$$

$$C_2H_5\text{—}CH(OH)\text{—}NH\text{-}CH_3 \rightarrow C_2H_5\text{-}CH{=}N\text{-}CH_3 + H_2O$$

In order to classify chemical reactions in a consistant manner which is suitable for chemical purposes, we define the following hierarchy of equivalent classes:

R-category ⊃ RA-type ⊃ RB-type ⊃ R1-type ⊃ R2-type ... etc.

1) Any two R-matrices R_1 and R_2 belong to the same R-category if their irreducible forms R^{o}_1 and R^{o}_2 differ only by row/column permutations.
2) Any two reactions belong to the same RA-type, if they belong to the same R-category, and if their non-zero entries refer to the same components of an associated vector of atoms.
3) Any two reactions of the same RA-type belong to the same RB-type, if they have the same positive entries

in the affected BE-matrices, where the R-matrices have non-zero entries.

4) Any two reactions of the same RB-type belong to the same R1-type, if those entries of the BE-matrices whose entries are affected by the R-matrix refer to the same components of the atom vector.
5) R2-, R3-, ...types may be defined in analogy to the R1-types by further classifying according to the second, third etc. sphere of neighboring atoms.

This hierarchic classification of chemical reactions by their R- and BE-matrices may not only serve as a means of formal ordering of reactions and as a basis of documentation systems, but can also serve as a device in the systematic computer-assisted deductive search for new chemical reactions, by an algorithm which finds all of the mathematically and chemically fitting pairs (B, E) of BE-matrices for a representation R-matrix of an R-category.

5. Geometric and Group-Theoretical Aspects of Constitutional Chemistry

The geometric and group theoretical aspects of the BE- and R-matrices are important for the solution of chemical problems.

Since these aspects have been discussed elsewhere[1)] in detail, a brief outline of the essential features will suffice here.

The BE- and the R-matrices belong to S(n), the additive group of all n x n symmetric matrices with integral entries. Let P_{ij} be an n x n matrix in which all entries are zero, except a "one" in the (i, j)-position.

Then

$$\{P_{ij} + P_{ji} \mid 1 \leq i \leq j \; n\} \cup \{P_{ii} \mid i = 1, \ldots, n\}$$

is a basis of the group S(n) whose rank is

$$\frac{n(n-1)}{2} + n = \frac{n(n+1)}{2}$$

We denote by Z^s the group whose elements are the lattice points of an euclidean s-dimensional space R^s, the group operation being vector addition. Then, Z^s, is a free abelian group of rank s with (1,0...0), (0,1,0...0) ...., (0,...,0,1) as a basis. Two free abelian groups are isomorphic if, and only if, they have the same rank. Therefore, Z^s and S(n) are isomorphic. The imbedding of Z^s in R^s is a geometric interpretation of S(n).

The BE-matrices are contained in $B(n) \leq S(n)$, the set of all n x n symmetric matrices with non-negative integral entries. The R-matrices belong to R(n) S(n), the group of all n x n symmetric integral matrices whose sum of entries is zero.

In the mapping $\pi : B(n) \rightarrow R^s, \pi|B(n)|$ is visualized as a cone in R^s with the vertex at the origin, and $\pi|R(n)|$ lies on a linear subspace going through the origin and having no other point in common with $\pi|B(n)|$.

The mapping $P: B(n) \rightarrow \{(b_{21} \ldots, b_{1n}; b_{21}, \ldots; \ldots; b_{n1}, \ldots, \ldots, \ldots, b_{nn})\}$ yields an imbedding of the BE-matrices B of an FIEM in R^{n^2}, an n^2-dimensional euclidean space.

The entries b_{ij} of B can be considered as the components of a vector in R^{n^2}, or also as the cartesian coordinates of a point P(B) in R^{n^2}. We call P(B) the BE-point of the BE-matrix B.

Similarly, an R-matrix R represents a vector in R^{n^2}. A chemical reaction which is represented by the transformation B + R = E can be geometrically interpreted by the vector R from the BE-point P(B) to the BE-point P(E).

The sum of the absolute values of the entries of R,

$$D(B,E) = \sum_{ij} |b_{ij} - e_{ij}| = \sum_{ij} |r_{ij}|$$

is equal to twice the number of valence electrons that participate in the reaction EM(B) → EM(E). We call D(B,E) the <u>chemical distance</u> between P(B) and P(E), or EM(B) and EM(E), respectively. Note that chemical distances refer to R^{n^2}, and to R^s.

The origin of the coordinate system in R^{n^2} corresponds to the zero matrix. Since the sum of entries $S = \sum_{ij} b_{ij}$, the number of valence electrons, is the same for all BE-matrices of an FIEM, and FIEM is mapped onto a lattice of points with non-negative integral coordinates in R^{n^2}, lying on a segment of a "hypershere surface" whose radius is S = D(B,O).

This remarkable topological order of the FIEM is not only theoretically appealing, but also didactically useful, because this order reveals well-defined universal logical structures in the immense wealth of individual facts.

In particular, the order provides a basis for the computer-oriented representation of molecular systems, and allows to formulate chemical facts and problems in a manner which is well-suited for their manipulation by computers.

6. MATCHEM

The mathematical model of constitutional chemistry which has been described in the preceeding sections can be used as a basis for a modular system of computer programs for the deductive solution of chemical problems.

Initially, the mathematical model was utilized for CICLOPS[3)4)], a pilot program for synthetic design.

The insight that the mathematical model may serve for the computer-assisted solution of a wide variety of chemical problems led to the design of MATCHEM, a modular system of computer programs, whose individual parts may be combined in different ways to suit different purposes. The original synthetic design program CICLOPS was modified via the intermediate stage MATSYN to yield the present synthetic design program EROS (Elaboration of Reactions for Organic Synthesis) which serves a similar purpose as SECS[6)] and the other reaction library oriented synthetic design programs LHASA[7)], SYNCHEM[8)], etc. In contrast to the pilot program CICLOPS, in EROS the modular type structure is emphasized more, in order to enable the combination with other parts of the MATCHEM system.

The synthetic design program EROS and its predecessors generate a tree of BE-matrices starting from one BE-matrix which refers to the synthetic target. This tree may be interpreted as a tree of synthetic pathways. If, however, the initial matrix does not represent a synthetic target and its by-products, but a given chemical compound in combination with a list of potential reactants, the generated tree of BE-matrices can be interpreted in terms of sequences of products obtainable from the initial reactants. In this form the program finds new uses for chemical compounds, e.g. industrial by-products, or permits the prediction of the conceivable fate of a chemical compound in the environment.

The mathematical model affords also entirely different approaches to chemical problems. If the initial and final ensemble of molecules, EM(B) and EM(E), of a chemical reaction, or a sequence of reactions are known, or attained by an educated guess, such as a comparison of substructures (see section 5.1.3) the difference of the corresponding BE-matrices E-B = R can be analyzed to yield the pathways of individual steps which lead from EM(B) to EM(E). These pathways may be sequences of intermediates in a reaction mechanism, or also sequences of synthetic intermediates. The precondition for this approach is that the indices of the atoms in EM(B) and EM(E) are appropriately assigned. The required assignment of atomic indices is discussed in section 5.3.1). All of these applications of the mathematical model, as well as the search of BE-matrix pairs (B, E) which fit a given R-matrix will be contained in MATCHEM.

In the now following sections some recently implemented parts of MATCHEM will be presented.

6.1 Manipulation of BE-Matrices

6.1.1 Canonical Ordering

A precondition for an efficient manipulation of BE-matrices in the computer is a canonical indexing of the atoms in a molecule. In order to generate a unique numbering we use the connectivity matrix of the molecular graph and the labels already assigned to its vertices, i.e. the chemical symbols.

An atom k is considered the j-th neighbor of i if j is the minimal number of bonds which separate i and k. All atoms k which meet this condition are called the j-th neighborhood of i. The zeroth neighborhood of i is atom i itself. We define an atomic descriptor as follows

$$d_i \equiv a_i | a_{11} | a_{12} \ldots | a_{22} \ldots a_{31} | a_{32}$$

where a_{jk} is the atomic number of atom k in the j-th neighborhood and all a_{jk} for the same j are put in descending order. The operator "|" means concatenation. The d_i are constructed in such a manner that all j-th neighborhoods are aligned. This is achieved by inserting dummy neighbors with atomic number of zero if necessary.

The sequencing of atoms is done by putting the atomic descriptors in descending order. Atoms which can not be assigned a unique number by this procedure are constitutionally equivalent. A final numbering of the atoms is reached by introducing arbitrary information. One atom is picked out of the highest ranking group of constitutionally equivalent atoms and its atomic number set higher than those of the other atoms. The atomic descriptors are modified accordingly and sorted again. The group of constitutionally equivalent atoms is thus split up. If a unique numbering is still not possible the modification of the atomic descriptors is reversed and another atom out of the now highest ranking group of equivalent atoms is picked and the procedure just described is repeated until all atoms are assigned a unique number.

It should be noted that it does not matter which atom is picked out of a group of atoms which cannot be assigned a unique number in the first place because these atoms are in fact constitutionally equivalent.

In this aspect our new indexing routine differs from our previous approach[4] . Our new procedure does not have the drawbacks of other methods[9] known, like instability of the calculations, oscillatory behaviour, nonconvergence and indeterminancy.

6.1.2 Direct Access to Structures and Substructures

The BE-matrix, together with its associated vectors (atoms, stereochemical parity bits) contains all structural information about a molecule. In a variety of applications, such information has to be stored and accessed rather frequently. Examples of such applications are documentation systems or the analysis of reaction graphs (trees or networks).

6.1.2.1 Inter-Machine Manipulation

The storage requirements of the full BE-matrix make it rather unsuitable for direct storage and retrieval. Compression can be achieved in two stages, depending upon the application. For data exchange between different computer systems (with possibly different codes and file structures), compression down to the level of characters (purely numerical) is a useful compromise. The compression method is straightforward in that it first assigns each structure (i.e. an intact molecule or a molecular fragment) a unique number, which may be derived from some sort of registry number, or may be generated by the system itself, in which case it usually will contain some graph information, such as node number or link pointer.

The information that is contained in the BE-matrix and its associated atom vector is compressed to the minimum number of bytes which is necessary for the expected range of values for each of these items.

Such a data format allows rather economical machine independent packing and unpacking of BE-matrices and is used for data transfer on sequentially organized, byte oriented media.

6.1.2.2 Intra-Machine Manipulation

Internal handling of large amounts of structural data puts more severe requirements on data format and on data organisation (in this paragraph, use of random access media, such as core memory banks - possibly with page or block organisation - or multi-track discs is assumed).

Structural information can be - and in fact is - further compressed by a.) replacing sequences of zero entries by pointers, b.) reducing the storage space for the remaining entries (atom symbol: 7 bit, diagonal elements: 4 bit, off diagonal upper triangle: 2 bit).

Further, the files contain references of different origin to other entries in that file. These appear in the form of numerical pointers and, if improperly organized, can use up large storage space. We developed a file organization to minimize storage requirement which makes use of relative pointers and pointer length classes. Furthermore, short access paths from pointing element (son) to pointed-at element (father), are postulated, i.e. minimization of core page-switching, or disk head movement respectively.

It has been found that the above two postulates lead to a data organization, whereby sons are assigned storage in the vicinity of their father. The address to be assigned is derived from the data themselves by a suitable hash-function, which assigns linearly increasing distances for cases without conflict, (addressed storage non-occupied), but switches to functions of second degree when conflicts (overflows) are to be handled. Since there are only moderate computing requirements, and since access times on the storage media determines processing speed, a rather moderate computer is needed for handling fairly large amounts of data. We are at present investigating the behaviour of a system with several thousand structures and their substructures (see 5.1.3) on a "mini" computer of the well-known PDP11 family that is equipped with moving head disk packs that have a storage capacity of about two mega byte each.

6.1.3 Hierarchically Organized Substructure Files

In a great many applications, a given structure is treated as a combine of its substructures. Depending on the context, substructures appear under the name of "functional group", "fragment", "reaction invariant" "building block" etc. Typical applications are optimization strategies in synthesis planning, avoidance of prohibited combinations of functional groups in generating chemical reactions, structure-activity correlations etc.

Computer generation, storage and retrieval of substructures is greatly facilitated, when the mathematical model of structural chemistry is employed to generate a hierarchically organized substructure file in a systematic fashion. Such hierarchical ordering is not only a prerequisite in avoiding duplication of information and missing links, it is also indispensable for setting up data organisations as described in the previous chapter (5.1.2).

The approach taken here treats the off diagonal entries of a BE-matrix as an ordered set of numbers. Each element in turn is lowered by one, the resulting matrix separated into blocks and the data file is tested for the presence of each resulting block. If the test is positive, then only a pointer to the generating structure ("father") is added to the file, and no further fragmentation is needed, since all successors must be in the file. If the test is negative, the son is entered into the file and its father (and brother if any) are pushed on a stack for further treatment. Each element of the stack is treated in the same way (thus becoming the father of the next generation) after the first son is processed. This depth-first fragmentation yields data-files that fulfil the requirement of the previous chapter, in that sons appear as close as possible to their father in the generation process, thereby leading to minimal pointer lengths.

With such a file structure queries take the form of paths within a directed graph and therefore are processed with a minimum of computing effort and extremely short response time.

6.2 Operating with R-Matrices

6.2.1 The Synthetic Design Program EROS

Our previous studies in developing CICLOPS, the prototype of a synthetic design program, have now led to a new stage.

We are now presenting EROS (Elaboration of Reactions for Organic Synthesis) which has matured to the point of becoming a routine tool for the synthetic chemist. A number of options allows the user to direct the program to meet his specific problems and needs.

Rather than taking the complete set of all mathematically possible reaction matrices a selection of only three R-matrices is contained in the program. These three R-categories include the majority of all known synthetic reactions and it is believed that most of the reactions which will be discovered in the future will also fall in these three R-categories. So the advantages of the mathematical model of providing also unprecedented reactions is largely retained. On the other hand substantial reduction of the output of unlikely intermediates is achieved.

EROS can be applied to the study of chemical reactions in two basically different ways: One can input several molecules and generate the products to be expected in the reactions of these molecules. Or one can look for all synthetic reactions lead to a target compound which is input.

Reactions mechanism can be studied by allowing certain electronic configurations for distinct atoms.

The reaction site is determined by stating which bonds are breakable. These can be found by a standard routine or preselected by the user. For each reaction the energy is calculated using parameters obtained from thermochemical data[10,11]. By defining energy limits, reactions can be rejected on thermodynamic grounds.

6.2.2 Predictor Systems

From the previous chapter it becomes obvious that R-matrices, when combined with suitable selection rules, are a general tool for predicting chemical reaction products.

In some applications, chemical selection rules may be less stringent because some other extraneous selection allows for further reduction of output to manageable size.

One group of applications are predictor systems. Such systems have, in general, access to structure files, i.e. files of intact molecules or molecular fragments of the nature described in chapters 5.1.2) and 5.1.3) whose entries are selected under a given aspect. Aspects may be toxicity, pharmaceutical activity, environmental impact, availability as an unwanted by-product in an industrial environment etc.

In such cases query entries are processed through reaction generators. These may be the complete set of R-categories, but of course a subset of those can be selected if the application justifies it. Output of the reaction generators is then checked against structure files and matches are output to the user.

One such system is at present being implemented under a research contract with a public agency with the purpose of detecting sources of environmentally active chemicals among chemicals regarded as harmless, themselves.

Although an essential part of such systems, the reaction generators, due to their algebraic nature, take up only a negligible amount of the total computing effort. It has therefore been remarked, not without justification, that in many applications predictor systems appear to the user as a mere enhancement of the capabilities of a documentation system. It should be stressed, however, that documentation systems will be able to support predictor modules only if they maintain the complete structural information in a form that is compatible with, or convertible to an algebraically defined BE-matrix. Design, or redesign of systems should take this fact into account.

6.3 Generation and Classification of R-Matrices

6.3.1 Determination of the Minimal Chemical Distance

Let EM(B) and EM(E) be the initial and final ensemble of molecules of a chemical reaction or sequence of reaction. This chemical conversion determines in a characteristic manner a correlation of the atoms in EM(E) and EM(B). We conjecture that chemical reactions proceed in such a manner that a minimum of valence electrons is relocated. This is criterion for the correlation of the atoms of EM(E) to the atoms of EM(B). We call the number of valence electrons which must be relocated during the transformation from EM(B) to EM(E) with a given correlation of atoms the chemical distance D(B, E) between these EM.

Mathematically the chemical distance D(B, E) can be formulated in the following way

Let (b_{ij}) be the BE-matrix of A and

(e_{ij}) be the BE-matrix of Z.

Then $$D(B,E) = \sum_{i,j=1,\ldots,n} |b_{ij} - e_{ij}|$$

For the chemically meaningful representation EM(B) → EM(E) the atoms of EM(B) must be assigned to atoms of EM(E) in such a manner that the chemical distance D has its minimal value.

This problem can be attacked in several ways

1) Exhaustive enumeration: all atoms with the same atomic number are permuted and the permutation corresponding to the minimal chemical distance is taken.

   Let EM(E) consist of n_1 atoms of atomic number O_1, n_2 atoms of atomic number O_2, ..., n_k atoms of atomic number O_k. Then one has to carry out $(n_1!) \cdot (n_2!) \ldots (n_k!)$ permutations. Since 10! = 3 628 800 this method is suitable only for very small molecules.

2) The criterion for minimization of the chemical distance can be partitioned into the three complementary criteria (i), (ii) and (iii).

   (i) Atoms with the same atomic number are given the numbers from the same set of numbers.

<u>Example I:</u> $\underset{3}{H_3}-\underset{2}{C}\equiv\underset{1}{N}: \rightarrow \underset{3}{H}-\underset{1}{N}\equiv\underset{2}{C}:$

In I (i) suffices to minimize the chemical distance.

(ii) The atoms are assigned in such a manner that, for atoms which are assigned to each other, as many spheres of neighbors as possible correspond.

This criterion follows from the assumption, that in a reaction each atom will try to maintain as many spheres of neighbors as possible.

Example II:

$$H_3C_5{-}C_3(C_4H_3)(C_6H_3){-}Cl_1 + H{-}O_2H \rightarrow H_3C_5{-}C_3(C_4H_3)(C_6H_3){-}O_2{-}H + H{-}Cl_1$$

In II (i) suffices to assign O and Cl; but for the assignment of the carbon-atoms (ii) is necessary.

(iii) If there are constitutionally symmetric molecules [4,12] in EM(B) and (or EM(E)) then an atom A in EM(B) can be assigned to more than one atom $Z_1, \ldots, Z_k$ in EM(E) even if one takes (ii) into account. Then there exist two possibilities:

1) A is the first atom in EM(B) which shall be assigned to an atom in EM(E). Then A can be assigned arbitrarily to one of the atoms $Z_1, \ldots Z_k$.

2) One has already assigned atoms $A_1 \rightarrow Z_1, \ldots, A_e \rightarrow Z_e$. Then A can be assigned to Z_i in such a way that the atoms $A_1, \ldots, A_e$ are in the same spheres of neighbors to A as the atoms $Z_1, \ldots, Z_e$ to Z_i.

This method will be iterated until all spheres of neighbors correspond to each other as good as possible.

Example III:

In example III one can find the optimal assignment only by criterion (iii).

In the computer program which we have developed for the minimization of the chemical distance we use algorithms from the field of operations research known as "optimal assignment"-methods.[13]

6.3.2 Documentation Systems for Chemical Reactions

Encoding and retrieval procedures for chemical structures have been devised and used since the times when structural organic chemistry became known. Although they had to be tailored to the technical nature of the data carrying media - be it spoken or printed words, manually processed card files, sequentially processed punch hole cards or magnetic tapes, - they all fulfilled their purpose at their time. The situation is rather different for chemical reactions. We still rely on such alchemistic means as naming-by-author or citing-by major-product etc. This is symptomatic for our state of knowledge. Apparently we have no systematic way of describing reactions until now. (The need, of course, was less pressing, since the number of reactions is much smaller than the number of structures.)

Approaches that use starting and/or final products as a vehicle for describing and classifying reactions must be regarded unsuitable in the light of our mathematical model: A reaction is a path between points on the subspace, that is defined by an FIEM. Although an individual path may be described by its starting, final (and possibly some intermediate points) equivalence classes of such paths cannot be adequately described and classified that way. What we need is a description of the path itself.

An R-matrix is such a description. In its most general form it contains no reference to a particular FIEM. Therefore (as explained in section 4.3) each irreducible R-matrix defines an R-category. The minimal reference to a specific FIEM, then, is through those elements of the atom vector, that correspond to the rows and columns in the irreducible R-matrix, in other words to the nonzero entries in the full R-matrix. Increasing degrees of specificity are obtained by a reference from this first set of atoms ("the reaction core") to their corresponding elements in the BE-matrix, that in turn enlarge the atom vector ("first sphere") etc.
A hierarchy is thereby achieved, that will lead to a reaction file structure, where each entry is referenced through the various levels of specificity up to the R-categories. Queries can be entered at any degree of generality.

In a practical system, a canonical ordering has to be imposed onto R-matrices, which is most easily achieved by selecting the lexicographically smallest irreducible R-matrix among the n! equivalent ones. Lexicographical priority is given to the diagonal elements. This, actually, is an arbitrary measure, but it is justified on the ground, that non-zero diagonal elements cause changes in valence state of the affected atom, thereby earmarking it as a key atom in the reaction core. Remaining symmetries in R-matrices, then, are resolved on the first occurence of differentiation in the course of increasing specificity.

It is worth remarking that the different R-categories govern widely different populations of known reactions. One obvious example is the simple four center reaction of the kind A-B + C-D = A-C + B-D, which is incidentally represented by the simplest R-matrix in closed shell chemistry that has a zero diagonal.

We have started the design of a model program to extract irreducible R-matrices from known reactions, to put them into canonical order, and to set the sequence of references in a hierarchical file, that is organized in a similar way as substructure files described in (5.1.3).

References

1. J. Dugundji, I. Ugi, Topics Curr. Chem. 39, 19 (1973) see also: I. Ugi, P. Gillespie, Angew. Chem. 83, 980, 982 (1971), Ibid. Internat. Edit. 10, 914, 915 (1971).

2. I. Ugi, P.D. Gillespie, C. Gillespie, Trans. N.Y. Acad. Sci. 34, 416 (1972).

3. J. Blair, J. Gasteiger, C. Gillespie, P. Gillespie, I. Ugi in "Computer Representation and Manipulation of Chemical Information", Ed. W.T. Wipke, S. Heller R. Feldmann, E. Hyde, Wiley, N.Y. (1974).

4. J. Blair, J. Gasteiger, C. Gillespie, P.D. Gillespie I. Ugi, Tetrahedron 30, 1845 (1974).

5. I. Ugi, J. Gasteiger, J. Brandt, J.F. Brunnert, W. Schubert, IBM-Nachr. 24, 185 (1974)
I. Ugi, IBM-Nachr. 24, 180 (1974).

6. W.T. Wipke, T.M. Dyott, J. Amer. Chem. Soc. 96, 4825 (1974), and earlier references.

7. E.J. Corey, W.J. Howe, H.W. Orf, D.A. Pensak, G. Petersson, J. Amer. Chem. Soc. 97, 6116 (1975), and earlier references.

8. H. Gelernter, N.S. Sridharan, H.J. Hart, S.C. Yen, F.W. Fowler, H.J. Shue, Topics Curr. Chem. 41, 113 (1973).

9. M. Randić, Journ. of Chem. Inform. and Comp. Sci., Vol. 15, No. 2, 1975 (105).

10. S.W. Benson, Thermochemical Kinetics, Wiley, N.Y. (1968), S.W. Benson et al. Chem. Rev. 69, 279 (1969).

11. T.L. Allen, J. Chem. Phys. 31, 1039 (1959), A.J. Kalb, A.L.M. Chung, T.L. Allen, J. Amer. Chem. Soc. 88, 2938 (1966).

12. J. Gasteiger, P. Gillespie, D. Marquarding, I. Ugi, Topics Curr. Chem. 48, 1 (1974).

13. R.E. Burkard, Methoden der ganzzahligen Optimierung, Springer-Verlag, Wien-New York (1972).

Acknowledgements

We acknowledge gratefully the financial support of or work by Deutsche Forschungsgemeinschaft and Fonds der Chemischen Industrie.

3

An Organic Chemist's View of Formal Languages

H. W. WHITLOCK, JR.

Dept. of Chemistry, University of Wisconsin, Madison, Wisc. 53706

It seems generally recognized that, except for the most difficult of cases, the introduction of mathematics tends to obscure problems rather than make their solution easier. Be that as it may we are quite intrigued by the relation between formal languages and chemical structures, particularly when considering the area of computerization of organic synthesis. The purpose of this paper is to outline some of our thoughts on the above subject, showing that one can apply common facts derived from language theory to questions of chemical interest. For the chemist readers we will define languages and grammars. We will then discuss a number of "theorems" of organic chemistry (formal proof will not be attempted). Finally, we will point out that a certain subset of organic synthesis, the Functional Group Switching Problem, is amenable to attack by viewing it as a problem in the context of formal languages.

Molecules as Strings of Symbols

As we will see, the tenets of language theory assume that one is dealing with strings of symbols: one dimensional linear assays. If we wish to analyze molecules according to this theory it behooves us to inquire as to what extent we may consider molecules to be linear entities. We immediately note that molecules, in particular the set of all structures containing rings, are inherently nonlinear in nature. On the other hand we recognize that we can _represent_ anything in a linear manner. If we make the rather interesting equation of representation with structure then it follows that structures, as represented, may be linear. Now one can carry this as far as one likes but it seems to the author that for most purposes one should restrict ones attention to linear representations that make a reasonable amount of intuitive chemical sense. For this reason we will not consider the case of cyclic structures.

What types of molecules can be naturally represented in a linear manner? Clearly straight-chained structures can. The representation of n-hexane,

$$CH_3CH_2CH_2CH_2CH_2CH_3,$$

is in fact what one thinks of when considering this molecule (representation equals structure). In particular this "structure" is intuitively a string of six symbols, namely

$$CH_3, CH_2, CH_2, CH_2, CH_2, CH_3.$$

Similarly ethyl crotonate is $CH_3CH{=}CHCOOCH_2CH_3$. This example points up two differences between a linear notation of a structure and the actual structure itself. The above is a string of symbols so we have to specify what symbols are involved. Possibilities for ethyl crotonate are: CH_3,CH,=,CH,CO,OCH_2,CH_3; CH=CH,COO,CH_2,CH_3; and CH_3,CH=CHCO,O,CH_2CH_3. Since in general one will attach meanings to the symbols involved,these different representations may have different meanings. For example, the last above might be described as sequentially a methyl, an αβ-unsaturated carbonyl, a dicoordinate oxygen, and an ethyl. This is a perfectly good definition of this molecule and, as a string, is somewhat more informative than a mere tabulation of the individual parts. The second major difference between linear notation and actual structures lies in the observation that strings have an inherent ordering from left to right while structures do not. This leads to a many into one mapping of ordinary line notations into structures.

Having seen that the ordinary line notation of unbranched structures may be more or less equated with the structure itself we next turn to the case of branched structures. Just as an unbranched structure corresponds to a string of symbols, a branched structure has as its counterpart a tree. The nodes of the tree are part structure symbols and the edges are bonds.* Thus 2,2,4-trimethylhexane may be represented by a number of trees, one of which is

---

*Multiple bonds may be represented several ways, for example as a nonstructural node.

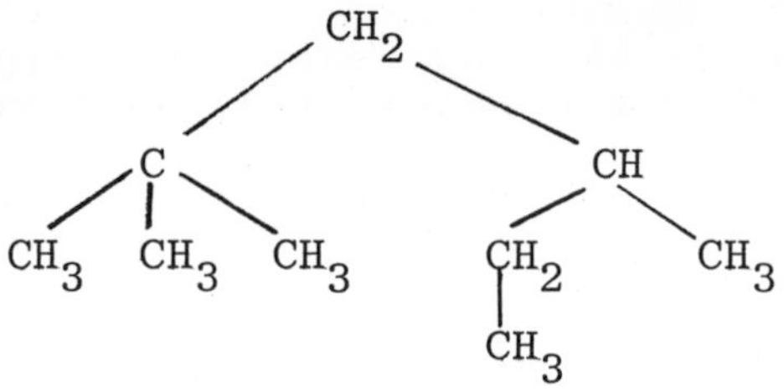

CH_2

C CH

CH_3 CH_3 CH_3 CH_2 CH_3

CH_3

One can define "global" part structures as was done above for ethyl crotonate and one notes that again there will be many tree representations per structure.

The point of this, of course, is that, as is well known(1), trees (in particular binary trees) may be represented as lists. The above tree has a list representation, $CH_2(C(CH_3)(CH_3)CH_3)CH(CH_2CH_3)CH_3$. Now this doesn't look too appealing to the chemists trained eye but $CH_3C(CH_3)(CH_3)CH_2CH(CH_3)CH_2CH_3$ does. This is a list representation of the tree

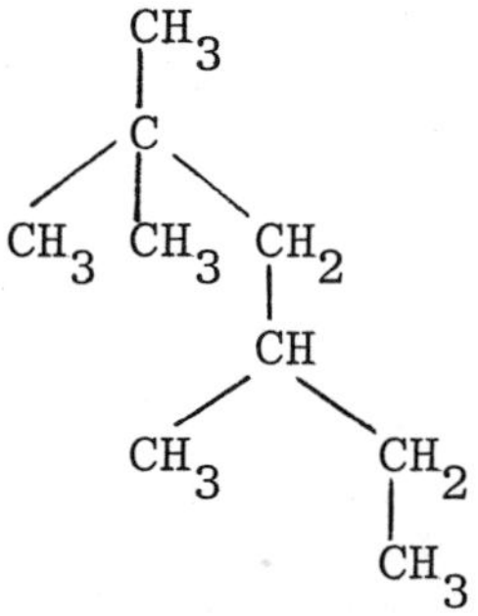

CH_3

C

CH_3 CH_3 CH_2

CH

CH_3 CH_2

CH_3

and is suspiciously close to our usual line notation of branched structures. We will show below that the set of all acyclic structures, wherein by "structure" we mean that intuitively well defined structural notation used by organic chemists, comprises a context free language. But first we must define the concept of languages and grammars.

Grammars and Languages

The following is a very brief introduction to the subject. We restrict ourselves to those aspects that are directly related to the problem of applying the theory of languages to organic chemistry. For a more complete introduction the reader is referred to a number of excellent texts.(2-5)

As conceived by Chomsky(6) the following are the

central aspects of this subject.

Symbols. There is some (normally finite) set of symbols from which strings (sentences) are made of. This set (V_T, the terminal vocabulary) would be {CH_3, CH_2} for the unbranched alkanes above and would be {CH_3, CH_2,CH,C,(,)} for the linear representation of branched alkanes.

As a grammar embodies the concept of derivation of some sentence in a language there is also defined a set of symbols (V_N, nonterminal vocabulary). These are used in derivations but do not appear in the final sentences of the language defined by the grammar of interest. The basic act of derivation involves replacement of a nonterminal symbol in a string by a string. For example the string $CH_3CH(R)R$ may be turned into the string $CH_3CH(CH_2$ R)R by replacing the nonterminal R by the string CH_2 R. On the other hand it might be changed into $CH_3CH(CH_3)R$ by replacing R by CH_3, depending on what our rules are for effecting these changes. Finally, there is some unique member of V_N, the "start" symbol, from which all sentences may be derived.

Productions. A production is just a rule for making the above changes. Replacement of R by CH_2R is symbolized by $R \rightarrow CH_2R$; replacement of R by CH_3, by $R \rightarrow CH_3$. As conventionally treated application of productions is permissive in the sense that there are no rules stating what production of some set must be applied to a given string. The problem of determining what series of productions will turn the unique start symbol S into some specified sentence then becomes an occasionally intricate puzzle. The general form of a production is $\alpha X \beta \rightarrow \alpha \omega \beta$, where X is some nonterminal symbol and α, β, and ω are arbitrary strings. Productions of this form with no restrictions on α, β, and ω are of type 0. Those with the restriction that ω not be the empty symbol are of type 1. An equivalent statement is that a type 1 production may be of the form $\alpha\gamma\beta \rightarrow \alpha\omega\beta$, length ($\omega$) $\geq$ length γ, γ being an arbitrary string containing at least one nonterminal symbol. The presence of the context α and β leads to the term context sensitive in describing these productions. Productions of the Form $X \rightarrow \omega$ are of type 2. The absence of a context, α and β leads to the term context free for this type of production. Finally, the simplest type of production, type 3 or regular, is restricted to be of either the form $X \rightarrow a$ (X in V_N, a in V_T) or $X \rightarrow aY$ (X and Y in V_N, a in V_T).

Note that all context sensitive productions are of type 0, all context free productions are context sensitive, and all regular productions are context free. Note also the direct analogy between productions as defined above and chemical reactions. The context sensitive production CH=CH CHOH → CH=CH CO is equally viewed as a reaction. Application of this production to the string $CH_3CHOHCH_2CH_2CH=CHCHOHCH_3$ may produce the new string $CH_3CHOHCH_2CH_2CH=CHOOCH_3$ but not $CH_3COCH_2CH_2CH=CHCHOHCH_3$. Clearly if structures are equated with strings, chemical reactions have as their counterpart productions. The term "context sensitive" has very similar meanings in both cases. Along these lines the production $CH(OCH_3)_2$ → CHO is of type 0 as it leads to a decrease in the length of the string.* The context free production CHO → CH(R)OH, corresponds to Grignard addition to an aldehyde, while examples of regular productions are $CH_2OH \rightarrow CH_3Br$, CHOH → CO, etc. Just as type 0 productions lead to a richer language the analogous chemical reactions lead to a richer more complex chemistry as we go from the simple regular "functional group switching" reactions to those involving blocking and deblocking reactions.

Grammar. A grammar is just a defined set of nonterminal and terminal symbols, a specified member of V_N (the start symbol) and a set of productions. Viewing reactions as productions we may define a chemistry as a set of molecular parts, a specified starting material, and a set of reactions. This assumes that the chemistry can be thrown into the proper grammatical form as discussed above.

Languages. For a defined grammar, its attendant language is the set of all strings (over V_T) that can be generated by repeated application of the grammar's productions, starting with the start symbol. Pursuing our analogy between grammars and synthesis, the language defined by some chemistry (chemical grammar) is the set of all molecules that can be synthesized from the specified starting material by repeated application of the reactions--a language of synthesizable structures for that chemistry. Just as grammars may be of type 0, 1, 2 or 3 according to the most complex type of production present, a language is of type 0 if specified by a type 0 grammar, etc. Note that while a grammar is

*Assuming our symbols are CH, OCH_3, (,), 2, CHO.

an exact (although sometimes opaque) definition of a language, a language does not in general specify a unique grammar. Chemical questions which are a direct transliteration of their corresponding language theory counterparts are: Given two chemical grammars: do they define the same language of synthesizable structures. Given a particular chemical grammar (chemistry) of say type 0, involving various blocking-deblocking sequences, is there a simpler chemistry that defines the same language of synthesizable structures. Given a chemistry and a particular molecule is the molecule a member of the chemistry's language (can it be synthesized?). Since this membership question becomes more and more complicated as we go from type 3 to type 0 languages, can we place some sort of upper and lower limits on the complexity of this problem within the context of language types? These questions will be dealt with below.

Examples

1) Grammar 1

$V_T = \{CH_3, CH_2, CH_4\}$
$V_N = \{ALKANE, R\}$
Start symbol = ALKANE

Productions:	ALKANE $\rightarrow$ CH_4	P1.1
	ALKANE $\rightarrow$ CH_3R	P1.2
	R $\rightarrow$ CH_3	P1.3
	R $\rightarrow$ CH_2R	P1.4

This is an example of a structural grammar. The productions correspond to rules for generating *n*-alkane structures rather than to chemical reactions. The language specified by this grammar is the set of all *n*-alkanes. Derivation of n-butane is achieved thusly:

$$ALKANE \xrightarrow{P1.2} CH_3R \xrightarrow{P1.4} CH_3CH_2R \xrightarrow{P1.4} CH_3CH_2CH_2R \xrightarrow{P1.3} CH_3CH_2CH_2CH_3$$

The sentence $CH_3CH(CH_3)CH_3$ is not in this language, nor are CH_3CH_2OH or CH_3R (this latter is a sentential form). Since all productions are of type 3 (X $\rightarrow$ a or X $\rightarrow$ aY) the set of all alkanes comprises a regular language. The membership question for regular languages is exceedingly simple. From the regular grammar above we may construct the algorithm or "machine" shown in Figure 1. Rules for constructing machines such as this from regular grammars are described elsewhere.(7) One starts in the start state and makes state

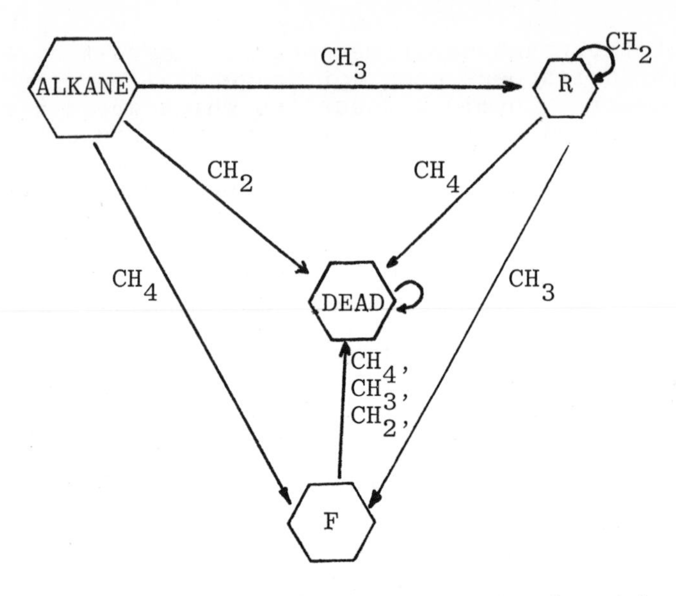

Figure 1. Finite state machine for recognizing members of the language defined by grammar 1. The start state is that labelled ALKANE. The accept state is that one labelled F.

transitions as one reads the string of interest from left to right. If one is in the accept state when no more symbols are left the string is a member of the language. If not, not. Note that the machine requires only a finite amount of memory; hence the term finite state machine.

2) GRAMMAR 2

V_T = {CH_3, CH_2, OH, MgBr, Br}

V_N = {S, OH, Br, MgBr}

Start State = S

Productions:

S → CH_3OH		P2.1
OH → Br		P2.2
Br → MgBr		P2.3
MgBr → CH_2OH		P2.4

This is a chemical grammar, the productions corresponding to reactions.* The language is the set of all 1-alkanols, 1-alkylmagnesium bromides and 1-bromoalkanes.

*The reader will note that V_N and V_T are not disjoint as they are supposed to be. This was done for clarity's sake as this grammar can be easily rewritten to conform to the standard definition.

Although the productions are not of all of the regular form $X \rightarrow a$ or $X \rightarrow aY$ it is easily rewritten to conform to this format. The language is thus a regular language and a finite state machine for recognizing members of the language is easily constructed. Note that a derivation of a sentence corresponds directly to its synthesis. Derivation of ethyl bromide proceeds as:

$$S \xrightarrow{P2.1} CH_3OH \xrightarrow{P2.2} CH_3Br \xrightarrow{P2.3} CH_3MgBr \xrightarrow{P2.4} CH_3CH_2OH$$

$$\xrightarrow{P2.2} CH_3CH_2Br$$

Theorem. The set of all wellformed acyclic structures comprises a deterministic context free language.

We have defined above what we mean by "structures," the ordinary line notation used by organic chemists wherein we rely on a string representation with branching indicated by parenthesization. The reader will recall that context free grammars allow more complicated nested productions, e.g. $S \rightarrow aSb$ than are allowed for by regular productions. Consider grammar 3 below:

GRAMMAR 3

$V_T = \{CH_3, CH_2, CH, CH_4, (,)\}$

$V_N = \{S, R\}$

Start Symbol = S

Productions:		
	$S \rightarrow CH_4$	P3.1
	$S \rightarrow CH_3R$	P3.2
	$R \rightarrow CH_3$	P3.3
	$R \rightarrow CH_2R$	P3.4
	$R \rightarrow CH(R)$	P3.5

This is just Grammar 1 with the addition of production P3.5. This production is not regular but is context free. It is this production that allows us to generate arbitrarily branched structures such as $CH_3CH(CH(CH_2CH_3)CH_3)CH_3$. That the language defined by this grammar is not regular (i.e. cannot be generated by a regular grammar) follows from its being homomorphically equivalent with the set of balanced parentheses. The recognition problem for context free languages is inherently more difficult than that for regular languages in that one needs unlimited memory (essentially for the reason in this case of remembering what branch one is currently examining). A complete grammar that handles a large subset of acyclic structures is presented in Figure 2. The language is not regular but is context free, as can be verified by observing the form of the productions. This grammar is the basis of a deterministic algorithm for parsing (i.e. recogniz-

V_T = {CH_4 CH_2O HCO_2H HCN CO_2 CO CH_3 OH COOH

CHO CN Cl Br HO HO_2C OHC NC CH_2 O

O_2C H_2C CH HC C = ≡ ( ) N}

V_N = {S LMV RMV CHN DB RDB TB RTB SM SRP

SLP DV LDV RDV TV LTV QV}

Start Symbol: S

PRODUCTIONS:

Structure

S → SM / SLP RMV / (LMV)2 TV RMV / (LMV)3 QV

RMV / LDV(RMV)2 / LDV RBD / TLV RTB /

LTV(RMV3 / QV(RMV)4 / (LMV3 TV / (LMV)4 QV /

(LMV)2 QV RDB

Left Monovalent

LMV → SLP / SLP CHN / (LMV)2 TV CHN / (LMV)2 TV /

(LMV)3 QV / (LMV)3 QV CHN / (LMV)2 QV DB/

LDV DB / LTV TB

Right Monovalent

RMV → SRP / DV RMV / TV(RMV)RMV) / TV(RMV)2 / TV

RDB / QV(RMV)2 RMV / QV(RMV)3 / QV RTB /

QV(RMV)RDB / (CHN)N RMV / QV(RMV)(RMV)RMV /

QV(RDB)RMV

Continued

Figure 2a. Context Free structural grammar for computer input of linear structure notation via teletype. Meanings of nonterminal symbols are as above.

Chaining

CHN → DV / DV CHN / TV(RMV) / TV(RMV)CHN / QV TB /

QV(RMV)(RMV) / QV(RMV)(RMV)CHN / QV(RMV)2

CHN / TV DB / QV(RMV)DB

Double Bond

DB → TV / = TV CHN / = QV(RMV) / = QB(RMV)CHN /

= QV DB

Right Double Bond

RDB → RDV / = TV RMV / = QV(RMV)RMV) / = QV(RMV)2

/ = QV RDB

Triple Bond

TB → ≡ QV / ≡ AV CHN

RTB → ≡ TV / ≡ QV RMV

SM → CH_4 / CH_2O / HCO_2H / HCN / CO_2 / CO

SRP → CH_3 / OH / CO_2H / CHO / CN / Cl / Br

SLP → CH_3 / HO / HO_2C / OHC / NC / Cl / Br

DV → CH_2 / O / CO / CO_2 / O_2C

LDV → CH_2 / H_2C

RDV → CH_2 / O / CO

TV → CH

LTV → CH / HC

QV → C

Figure 2a. Continued

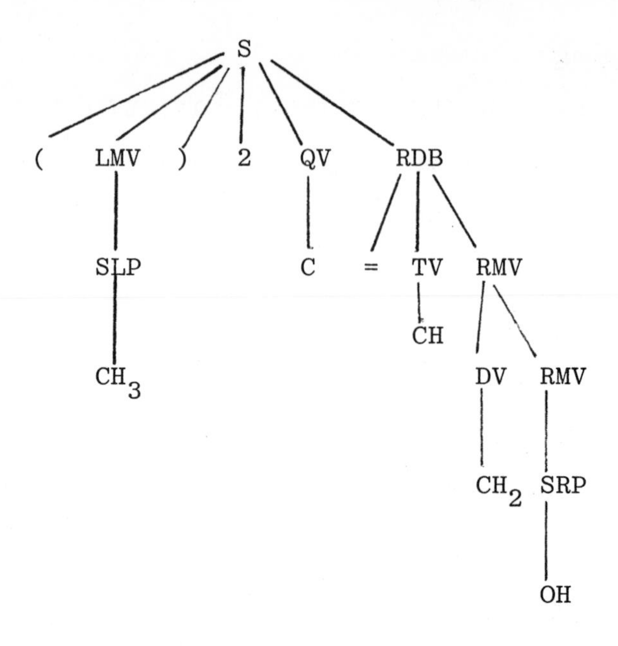

Figure 2b. Parse tree of $(CH_3)_2C{=}CHCH_2OH$, a representative terpenoid

ing) structures input into a LISP organic synthesis program written by P. Blower (8). This grammar generates most common functional groups (those not included such as $-NO_2$ were left out for nongrammatical reasons) and includes chaining, the representation of repeated subunits by the enclosing of them within brackets. The representations of n-butane, $CH_3CH_2CH_2CH_3$, $CH_3(CH_2)_2CH_3$, $C_2H_5C_2H_5$, $CH_3CH_2C_2H_5$, and others are all accepted by this program, parsed, and converted into internal representations, pointing up the fact that there is a many into one mapping of structural representations into structure.

We have seen above a simple synthetic grammar wherein the productions are directly derived from chemical reactions and the resulting language of synthesizable structures is the set of all molecules that can be synthesized from some specified precursor by application of the reactions. In the simple case of 1-alkanols and 1-bromoalkanes the recognition problem is trivial--we can answer it in a time proportional to the length of the molecule. We are naturally curious as to what is the minimally complex grammar needed to mimic organic synthesis of the more conventionally complex type. This is not a silly question for the following two reasons. Although one can quibble as to the extent to which acyclic molecules may be equated with strings there is no question that the similarity is marked. Moreover, at least from a complexity sense it is clear that algorithms for recognizing regular structures are inherently simpler than those for recognizing context free ones. Similarly the recognition of members of context sensitive languages is more difficult yet. This is true regardless of whether one is talking about the recognition of well formed or synthesizable structures, or whether one is doing this in a formal mathematical sense or via computer programs. Secondly it is not obvious that the question of synthesizability is in fact answerable at all for all well defined organic molecules. By answerable we mean having a recognition procedure that will terminate in some (not necessarily short) period of time with the answer yes or no. The set of context sensitive languages are recognized by the socalled linear bounded automata and the question of membership of some string in a defined context sensitive language is known to be answerable. Type 0 languages on the other hand are recognized by Turing machines and the question of membership is <u>not</u> answerable for type 0 languages as a set, although it may be for some subset. Thus if we cannot develop cogent arguments for the sufficiency of context sensitive languages as a model for oganic synthesis we are left with the possibility that organic synthesis cannot be "solved" by computer. No guarantees though, since organic chemistry is not a closed science and what may be an acceptible synthesis under some circumstances will be unacceptible under others. We first show by a counterexample that context free languages are insufficient model for organic synthesis. We then argue (alas we cannot prove, for the above reasons) that context sensitive languages <u>are</u> a sufficient model.

<u>Theorem</u>. There exists a language of synthesizable structures that is not context free.

It is well known that languages such as ww, where w is some string over V_T, are not context free. For example the set of all alkanes $CH_3CH(R_1)R_2$ where $R_1=R_2$ is not a context free language. The set of all tertiary alcohols derived by Grignard addition to methyl acetate, represented as $CH_3COH(R_1)R_2$, $R_1=R_2$ is thus not context free. That it is context sensitive follows from construction of a context sensitive grammar for generating this set (grammar 4, Figure 3).

GRAMMAR 4

$V_T = \{CH_3, CH, OH, CH_2, (,)\}$

$V_N = \{S, R, \Delta, \nabla, \#, X, Y, Z, \{W_i | i \varepsilon \{CH_3, CH_2, CH (,)\}\}\}$

Start symbol = S

$S \rightarrow CH_3CHOH \; \Delta \; \nabla \; R \; \#$

$R \rightarrow CH_3X | CH_2R | CH(R)R$

$iX \rightarrow Xi, \; i \varepsilon \{CH_3, CH_2, CH, (,)\}$

$RX \rightarrow R$

$XX \rightarrow X$

$\nabla X \rightarrow \nabla Y$

$Yi \rightarrow iY$

$YR \rightarrow R$

$YX \rightarrow X$

$Y\# \rightarrow Z$

$iZ \rightarrow W_iZi$

$jW_i \rightarrow W_ij, \; j \varepsilon \{CH_3, CH_2, CH, (,)\}$

$\nabla W_i \rightarrow W_i \nabla$

$\Delta W_i \rightarrow \Delta i$

$\Delta \rightarrow ($

$\nabla \rightarrow)$

$Z \rightarrow$

Language = $CH_3CHOH(R_1)R_2, \; R_1 = R_2$

Figure 3

Now if we consider further the relationship between derivation of a sentence and its synthesis, especially the relationship between the intermediate sentential forms and the precursor structures in the syntheses we are led to the following theorem.

Theorem (?). For a chemistry derived from functional group switching reactions, condensation reactions, demasking of masked functional groups, and blocking and deblocking reactions: the language so defined is a context sensitive one (i.e. there exists an equivalent context sensitive grammar that generates the same language). The question of synthesizability is thus answerable.

This follows from the socalled workspace theorem of context sensitive languages.(9) The "proof" of this simply entails stating informally the proof of the workspace theorem, letting synthetic intermediates correspond to individual steps in the derivation of the target molecule (sentence).

1) Consider a synthesis D of the form:

$$S \rightarrow X_o \rightarrow X_1 \rightarrow \cdots X_n = \text{target}$$

where S is the starting material (symbol), X_n is a

sentence in the language of synthesizable structures and $X_i \rightarrow X_{i+1}$ is some conversion. Assume moreover that there is some estimate SIZE (X_i) of the size of each X_i. We define the complexity $C(X_n,D)$ of the target X_n for this particular synthesis to the $\max\{SIZE(X_i), 0 \leq i \leq n\}$, the size of the largest compound involved in the synthesis. For a reverse salami synthesis, $C(X_n)$ is just the size of X_n. If the last step of the synthesis involves a deblocking reaction, $C(X_n,D) = SIZE(X_{n-1})$. $C(X_n,D)$ is thus a measure of the complexity of a particular synthesis (D) of a particular member X_n of the language.

2) Considering all possible syntheses of X_n, the workspace of X_n, $WS(X_n)$, is $\min\{C(X_n,D_m) | D_m$ is some synthesis of $X_n\}$. $WS(X_n)$ is thus a measure of the complexity of the least complex synthesis of X_n. While some syntheses may be arbitrarily complex it's the least complex synthesis that is important. In particular the ratio $WS(X_n)/SIZE(X_n)$ is a measure of the difficulty of the synthesis of X_n.

3) The workspace theorem simply states that if for some arbitrary type 0 grammar there is some one number p such that

$$\frac{WS(X)}{SIZE(X)} \leq p$$

for all X's in the language, the language is context sensitive and thus there exists an equivalent context sensitive grammar.

The argument for the existence of some p is as follows:

a) If we use functional group switching reactions as discussed below, p is obviously unity.

b) Condensation reactions will have a p greater than unity. Consider for example for the synthetic problem

$$CH_3OH \xrightarrow{?} CH_3CH_2COOH$$

Assume that the least complex route is

$$CH_3OH \longrightarrow CH_3Br \longrightarrow CH_3CH(COOCH_3)_2 \longrightarrow CH_3CH_2COOH$$

SIZE	2	3	8	3

The ratio $WS(X_n)/SIZE(X_n) = 2.7$. However as the molecules get larger, assuming that the condensing agents (e.g. malonic ester) stay constant in size, this ratio decreases. Thus for

$$CH_3CH_2OH \rightarrow CH_3CH_2Br \rightarrow CH_3CH_2CH(COOCH_3)_2 \longrightarrow CH_3(CH_2)_2COOH$$

SIZE 3 3 9 4

the ratio decreases to 2.25. For syntheses involving condensation reactions it would seem that there will exist some p.

c) The same argument may be applied to blocking-deblocking sequences and to operation involving masked functional groups. It seems inevitible that if we consider blocking groups to incrementally increase the complexity of some precursor the ratio $WS(X_n)/SIZE(X_n)$ must approach some number as $size(X_n)$ gets larger. In fact the only siutation wherein this would not be the case would seem to be one wherein the complexity of some necessary blocking group depended exponentially on the complexity of the intermediate being blocked. This type of exponential blocking group is unknown to organic chemistry and indeed seems foreign to the very concept of isolated and interacting functional groups. Functional groups, even in a relaxed definition are inherently local affairs.

d) Consideration of a number of published syntheses of complex natural products suggests a p of approximately 1.8, a remarkably low number. The author finds it striking indeed that those syntheses involving the selective reagents characteristic of "synthetic methods" chemistry so much in the vogue lately seem to have a smaller workspace than those carried out in the grand traditional manner.

We conclude that the workspace theorem is probably valid for organic synthesis, although it is certainly true that the above analysis ignores questions dealing with the formal relationship between synthetic schemes and derivations.

It would be nice to be able to write a program that would take as input two sets of chemical reactions, the second being the first augmented with some new reagent, and give as output the answer to the question: "Does this new synthetic method allow us to do anything we couldn't do in its absence?". This chemical critique program would be at least useful to editors of chemical journals. We note alas that for the set of context sensitive languages the problem of equivalence of two grammars is not answerable. Thus the above undertaking would seem to be a dubious one. Now of course the fact that the equivalence problem is unanswerable for the set of all context sensitive languages does not mean that it is so for some subset, for example augmenting a chemistry containing sodium hydroxide by addition of the reagent potassium hydroxide. On the other hand we note that the equivalency problem is not answerable for even the simpler set of context free languages so it seems unlikely that one could write a general program that would compare two chemistrys based on context sensitive reactions and that would halt in some finite time with the equivalence answer.

The Functional Group Switching Problem. (10)

One frequently has occasion when devising a synthetic scheme to adjust functionality in a molecule in a manner that is only indirectly related to the synthetic problem at hand. For example if one desires the transformation

$COOC_2H_5$ → $C(CH_3)_2OH$

it is necessary to block the ketone toward action of the organometallic reagent employed. Bearing in mind the assumptions built into this problem we may define a "molecule" as being simply an ordered set of functional groups. Chemically this is equivalent to the idea of a molecule's being a set of functional groups imbedded in a static molecular framework.

Definition of a reaction as an ordered triplet, (reagent, precursor functional group, product functional group) then corresponds to the assumption that reactions only interconvert functional groups; they do not lead to increments of the carbon skeleton, or if they do, it is only to finite and limited degree. Now this sounds like a rather restricted picture of organic chemistry, and it is, but it is surprisingly close to the way one thinks about blocking group interconversions. We can pose the functional group switching problem within the context of this structural notation in the following manner. What is the shortest sequence of reagents that will transform a defined starting material $S = (S_1, S_2, \cdots S_n)$, where S_i is the ith functional group of compound S, into the target $T = (T_1, T_2, \cdots T_n)$. The reagent sequence turns S_1 into T_1, S_2 into T_2. etc. With this definition of the functional group switching problem we are lead to the following.

Theorem. The set of all sequences of reagents that affects a stated functional group switching problem comprises a regular language.

The proof of this theorem is fairly obvious and not too interesting. What is interesting is the consequence of the theorem. The proof follows from the recognition that our functional group reaction dictionary is a finite directed graph wherein the nodes are labelled with functional groups and the edges with reagents. A small reaction graph is shown in Figure 4. This is of the same form as the finite state machine in Figure 1 and if we define a start state (e.g. CO) and an accept state (e.g. CHOAc) it is a finite state machine for recognizing all sequences of reagents that will turn a ketone into a secondary acetate. The reagent sequence ($NaBH_4$ DHP H_3O NaOH Ac_2O) is a member of the language so defined while ($NaBH_4$ DHP H_3O CrO_3/py Ac_2O) is not. The relationship between regular languages, regular grammars, and finite state machines is such that most interesting questions dealing with them are answerable. Whether a particular string is in the language is answerable (i.e. does this reagent sequence do the trick). More interesting however is the following which represents a general solution to the functional group switching problem. Our reaction dictionary defines, for the functional group switching problem $S_1 \rightarrow T_1$ (S_1 and T_1 single functional groups) a language L_1 of sufficient synthetic sequences. For a problem $S_2 \rightarrow T_2$ a language L_2 is defined. If we want to convert the binary compound (S_1 S_2) into the compound (T_1 T_2), the language for this is the intersection of L_1 and L_2, i.e. those members common to L_1 and L_2 are sequences that convert S_1 into T_1 and S_2 into T_2. It is known that the intersection of two regular languages is itself regular so the problem of finding the shortest sequence of reagents for effecting (S_1 S_2) $\xrightarrow{?}$ (T_1 T_2) is that of finding the shortest string in a regular language. The details of constructing the algorithm are presented elsewhere (10) but this points up what the author feels to be one of the really pretty aspects of language theory. The conventional proof of the statement that the membership question is answerable for regular languages is constructive in that a proveably correct algorithm for doing so is developed. One may then start from this fact and develop more efficient ways of achieving this end. We offer as evidence that the linguistic approach to

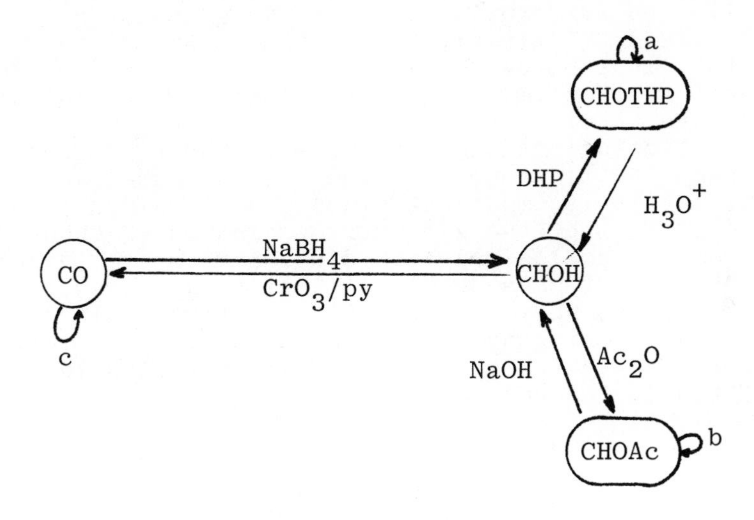

a ε\{DHP, $NABH_4$, CrO_3Py, Ac_2O, NaOH\}

b ε\{DHP, $NaBH_4$, CrO_3Py, Ac_2O, H_3O\}

c ε\{DHP, CrO_3py, Ac_2O, H_3O, NaOH\}

Figure 4. A small reaction graph for the four functional groups CO (ketone), CHOH (secondary alcohol), CHOTHP (tetrahydropyranyl ether), and CHOAc (secondary acetate)

organic synthesis is of more than just idle interest by presenting in Figure 5 some representative problems with their solutions. These examples, although presented in the rather disembodied order n-tuplet structural notation, clearly show that the results of this approach represent nonobvious answers to nontrivial problems.

It would seem that the basic feature of this approach to organic synthesis is applicable to more complicated structural models as long as several conditions are met. Firstly the synthetic problem of interest must be capable of dissection into some independent subproblems. This is so because the solution procedure involves at least implicit construction of the part solutions and working with their intersection. Secondly the finiteness of the various sub-reaction dictionaries is important since one is guided in solving a problem by the exhaustive solution of subproblems. For real molecules reaction dictionaries are infinite. Application of ones chemists' intiution suggests that only a finite part of a reaction dictionary is chemically interesting however. While there is an infinite number of structures that can be involved as intermediates in the conversion

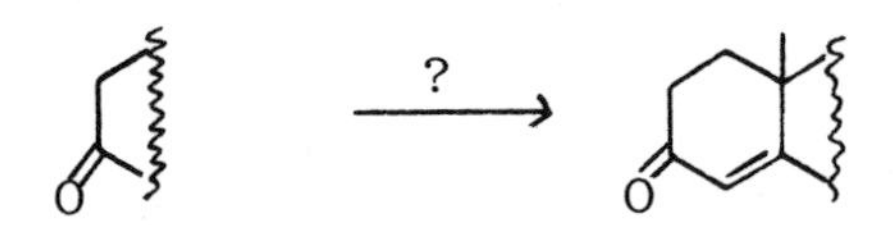

only a small number are realistic in nature when one has this particular end in mind. One problem of immediate concern is that of automating the making finite of the reaction dictionary for functionalities such as ketones that are destined for annulation. Interaction of the computer with the chemist is clearly necessary in this respect.

A related condition deals with the very state nature of this approach. The problem RCH=CH-COR $\xrightarrow{?}$ RCH=CH-CHOHR is not properly viewed as (CH=CH, CO) $\xrightarrow{?}$ (CH=CH, CHOH) since this entails the incorrect assumption that one is dealing with an isolated double bond and ketone. The intersection of the two reaction dictionaries would incorrectly have "no reaction" for the reagent $(CH_3)_2CuLi$. One thus must treat interacting functional groups as larger entities. This is an acceptable price to pay except for two consequences. Reaction dictionaries of aggregate functional groups can be very large and their generation by hand is a tedious and time consuming affair. It appears worth-

START: [RCHOEE RCO COOH]

TARGET: [RCHOEE RCO RCHOH]

SEQUENCE: (Glycol, EVE, RLi, $NaBH_4$, Ac_2O, H_3O, EVE, OH)

START: [CH_2OH RCO COOH]

TARGET: [CH_2OH RCO R_2COH]

SEQUENCE: (GLYCOL, RLi RMgX H_3O)

START: [CH_2OH CH_2OAc COOMe CH_2OEE]

TARGET: [CH_2OAc CH_2OH COOMe CH_2Br]

SEQUENCE: (CrO_3/py H_3O TsCl NaBr OH EVE $NaBH_4$ Ac_2O CH_2N_2 H_3O)

START: [CH_2OH CH_2OAc COOMe CH_2OEE]

TARGET: [CH_2OH CH_2OAc COOMe CH_2Br]

SEQUENCE: (CrO_3/py H_3O TsCl NaBr $NaBH_4$)

Figure 5. Functional Group Switching problems solved as in Ref. 10. RCHOEE is the ethoxyethyl ether of a secondary alcohol; EVE is ethylvinyl ether.

while to solve this reaction dictionary generation problem by a combination of chemist and computer with the chemist exercising his judgement in paring the growth of the reaction dictionary and making rather delicate value judgements on exactly what reaction is expected of some aggregate functional group under some set of reaction conditions.(11)

Literatue Cited

(1) McCarthy, J. Abrahams, P. W., Edwards, D. J., Hart, T. P., Levin, M. I., "LISP 1.5 Programmer's Mannual", MIT Press (1965).
(2) Hopcroft, J. E., and Ullman, J. D., "Formal Languages and Their Relation to Automata", Addison-Wesley, Reading, MA, 1969.
(3) Ginsburg, S., "The Mathematical Theory of Context Free Languages", McGraw-Hill, New York, 1966.
(4) Kain, R. Y., "Automata Theory: Machines and Languages", McGraw-Hill, New York, 1972.
(5) Aho, A. V., and Ullman, J. D., "The Theory of Parsing, Translation and Compiling", two volumes, Prentice Hall, Englewood Cliffs, N.J., 1973.
(6) Chomsky, N., Handbook of Math. Psych. (19), 2 Wiley, New York, pp. 323-418.
(7) Hennie, F. C., "Finite-State Models for Logical Machines" Wiley, New York, 1968.
(8) Blower, P., Ph.D. Thesis, University of Wisconsin-Madison, 1975.
(9) Salomaa, A., "Formal Languages", Academic Press, New York, 1973.
(10) This subject is discussed at lenght in, Whitlock, H. W., Jr., J. Am. Chem. Soc. (1976), 98, 0000.
(11) Support of this work by the National Science Foundation is gratefully acknowledged.

4

A Chemical Engineering View of Reaction Path Synthesis

RAKESH GOVIND and GARY J. POWERS

Dept. of Chemical Engineering, Carnegie-Mellon Univ., Pittsburgh, Pa. 15213

The synthesis, analysis, and evaluation of reaction paths is of fundamental importance in the chemical process industries. Research and development chemists and chemical engineers are often confronted with problems which are best solved by utilizing chemical reactions. Figure 1 shows a matrix of the common problems and possible solutions encountered in the chemical process industries. Problems associated with products may be classified into those which deal with compounds or mixtures of compounds which are a new market for the company and those which are existing products. Separation problems which might be solved by chemical reactions (e.g. use chemical reactions to convert one or more of the species in a mixture

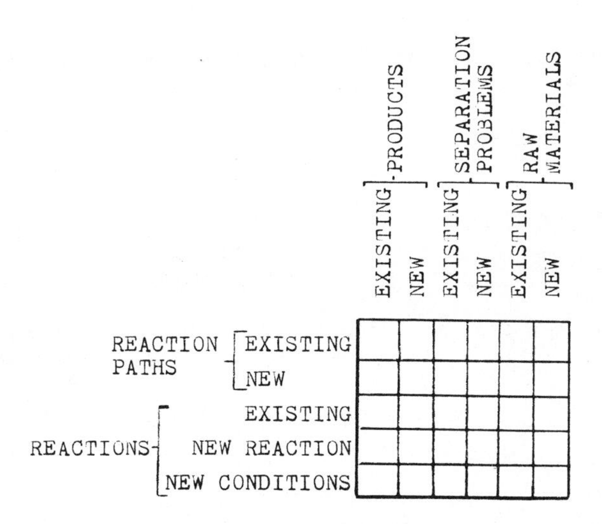

Figure 1. Problems commonly encountered in the chemical process industries which could be solved by use of reactions or reaction paths

to species which are more easily separated or need not be separated at all) may also be classified by whether they are new problems or existing ones. New problems arise with new products or new legislation related to pollution or product purity. Problems may also arise due to changing availability of raw materials. For example, a new reaction path may make an existing process and associated raw materials highly available. The problem is to find new uses for the raw materials and the process. In addition, it might be desirable to develop reaction paths which utilize (or avoid) certain technology, patents, etc.

Each of these problems may be attacked by inventing new reaction paths. The paths might be composed of known reaction steps in a new combination. Or, a new reaction step could be developed and utilized in a sequence of reactions. The new reaction could be a reaction of functional groups which was previously unknown or a previously known reaction which is made to go under new conditions.

Evaluation of Industrial Reactions

The evaluation of industrial reaction paths depends on a detailed understanding of the costs associated with each reaction step. A simple cost function is

Cost of Reaction Path = f(Overall Raw Material Cost, Utility Costs, Equipment Costs, Safety, Reliability, Flexibility) (1)

Raw Material Costs = g(Raw Material Cost/mole, Stoichiometry, Yield) (2)

Utility Costs = h(Thermochemistry, Reaction Conditions, Separation Difficulty) (3)

Separation Difficulty = i(Property Differences, Concentrations, Recovery, Separation Conditions) (4)

Equipment Costs = j(Reaction Kinetics, Separation Difficulty, Corrosion, Conditions) (5)

Safety = k(Species Properties, Conditions, Other Possible Reactions) (6)

Reliability = ℓ(Knowledge of Reaction, Separation, etc.) (7)

Flexibility = m(Range of Operating Conditions) (8)

Obviously a great deal of information is required to accurately compute the value of this function. One of the major functions of an industrial research and development effort is to generate the knowledge and data necessary for decision-making relative to each reaction step and path.(1)

Industrial Reaction Path Synthesis

The challenge in systematic reaction path synthesis is to generate reaction paths and steps which could be viable alternatives in an industrial environment. The reaction paths used in industrial environments have the following characteristics:

1. Small Target Molecules. The basic petrochemical and fine chemical industries commonly deal with molecules with fewer than twenty hetero- and carbon-atoms.
2. Multiple Target Molecules (Reaction Networks). The chemical industry is a network of reactions fed by 3 to 5 basic materials and producing hundreds of target molecules. The paths to each target molecule interact with each other by sharing raw materials and byproducts. See Figure 2 for part of a reaction network.
3. Stoichiometry Must Be Known. The stoichiometry for the path must be known in detail. This means that all main reaction products must be considered at each reaction step. The fact that 2 moles of NaCl or other simple reaction products are produced during a reaction can have a major impact on the reaction's economics.
4. Yield. The fraction of the limiting reagent which is transformed into the desired target molecule must be known.
5. Byproducts. The amount and type of byproduct produced are necessary pieces of information. The main reaction byproducts and side reaction byproducts are needed for both the yield calculation and the determination of separation difficulty, corrosiveness, safety, etc.
6. Impurity Reactions. The large-scale production of molecules demands a knowledge of the fate of impurities which enter with the raw materials or are produced in the reaction steps. These impurities can cause considerable pollution and product quality problems. In many processes more equipment and energy is devoted to controlling the impurities than the main reagents and products. Figure 3 illustrates the types of reactions which could occur between impurities and other species in the reaction mixture.

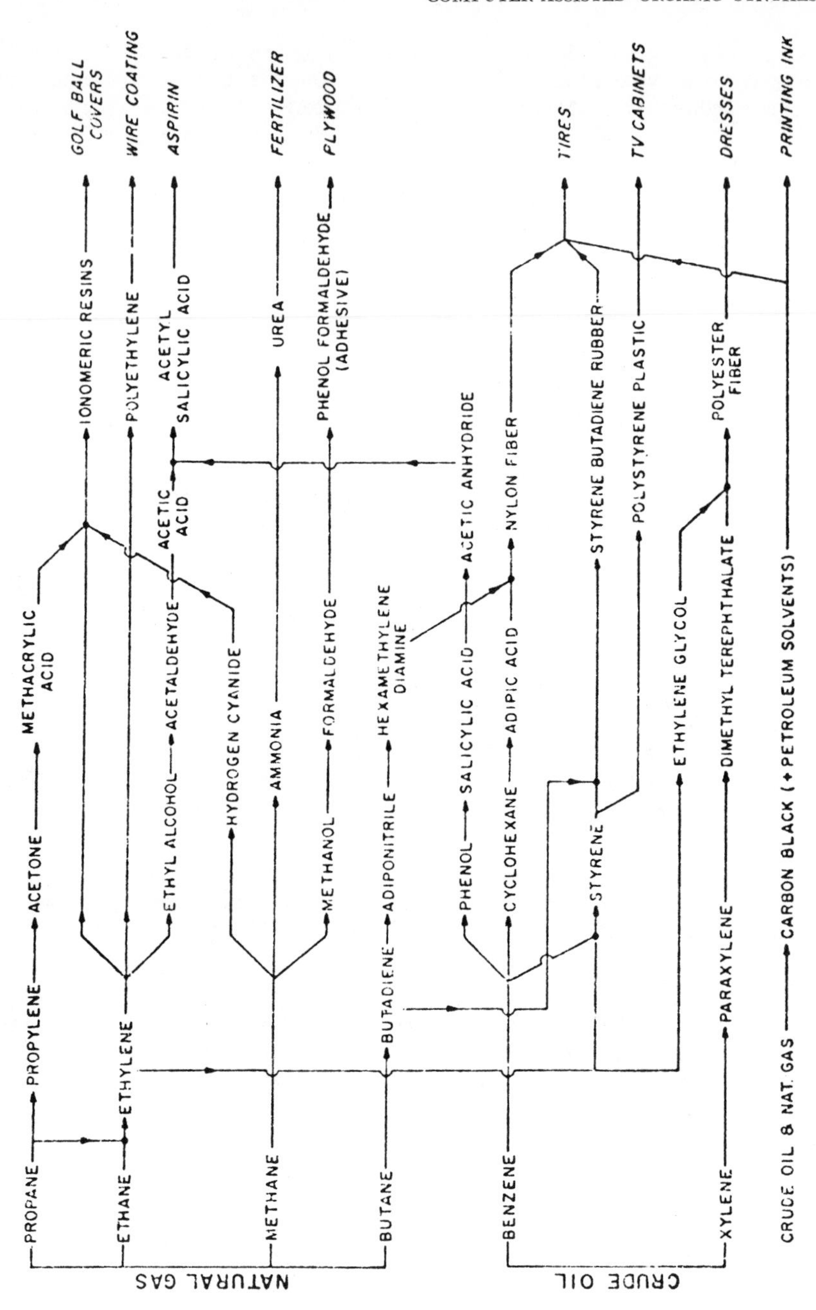

Figure 2. A network of reactions for part of the petrochemical industry

MAIN REACTIONS AT OTHER SITES

PARALLEL $R_1 + R_2 = P$: $R_1 + R_2 = BP_1$

SERIAL $R_1 + P = BP_2$: $R_1 + BP_1 = BP_3$

OTHER REACTIONS THAT OCCUR AT THE SAME CONDITIONS

$R_1 + R_2$ $R_1 + R_1$ $R_1 + P$

$R_2 + R_2$ $R_2 + P$ $R_1 + BP_1$

$P + P$ $P + BP_1$ etc.

IMPURITIES IN REAGENTS

$I_1 + R_1$ $I_1 + R_2$ $I_1 + P$

$I_1 + I_1$ $I_1 + I_2$ $I_1 + BP_1$

$I_2 + BP_1$ etc.

Figure 3. Reactions which could occur between the species (R ≡ reagents, P ≡ products, I ≡ impurities, BP ≡ byproducts) in a reaction mixture

7. Reaction Conditions. All of the features discussed above depend on the reaction conditions. It is necessary to have information on

   a. Phase(s) in which the reaction takes place.
   b. Solvents required (the fewer the better).
   c. Catalysts
   d. Temperature
   e. Pressure
   f. Concentrations
   g. Time (mixing)

Of course, this is asking for a great deal of information. Can modern theories and applications of reaction path synthesis make any contribution to this problem? Are the industrially important molecules too "simple" for current reaction path synthesis programs? Is too much specific data required?

In the following sections we present a brief review of work in computer-assisted reaction path synthesis. The work is compared with the needs of industrial reaction-path problems. Finally, a program called REACT which we have developed for reaction path synthesis is described and illustrated.

Representation of Molecules and Reactions

One key element in developing a reaction path program is the selection of representations of molecules and their reactions which are appropriate for a given problem. Figure 4 conceptually compares the predictive power and generality of three representations. Predictive power is defined as the ability to predict the conditions, stoichiometry, byproducts, yield, etc., of a given reaction step in each representation. Generality is the ability to generate "all" possible reactions.

The three representations are shown schematically in Figure 5. The most general types of representation use a mathematical abstraction of reaction which consider all the possible means for making and breaking bonds between atoms. Hendrickson (2) has investigated a general representation in which molecules are represented by the "character" of the carbon sites which occur within the molecule. The character of a site is a two digit number which indicates whether a carbon atom is attached to σ other carbon atoms by sigma bonds or to "functional" groups. The functionality of a carbon site is defined as the number of π bonds and heteroatoms attached to the carbon. The character of a site is given by

$$c = 10\sigma + f \tag{9}$$

where

$$f = \pi + z \tag{10}$$

Valence constraints give

$$\sigma + \pi + h + z \leq 4 \tag{11}$$

where h is the number of bonds to H or less electronegative atoms.

With this definition of carbon site, Hendrickson was able to classify all possible carbon sites and the "reactions" which might interconnect them. Figure 6 gives the reaction triangle for this representation.

Camp and Powers extended this representation by allowing the functionality (f) to be separated into π and z dimensions. The character of a site then becomes

$$c = 100\sigma + 10\ f + z \tag{12}$$

In addition, Camp (3) developed the site characteristics for heteroatoms. These extensions transform Hendrickson's triangle

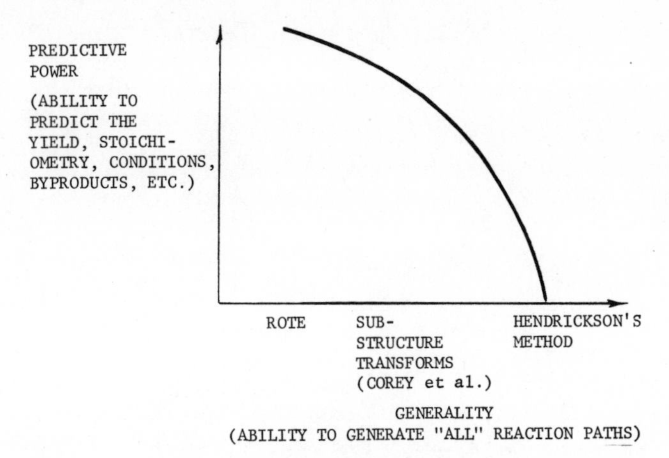

Figure 4. Ability to predict performance vs. generality for three representations of reaction

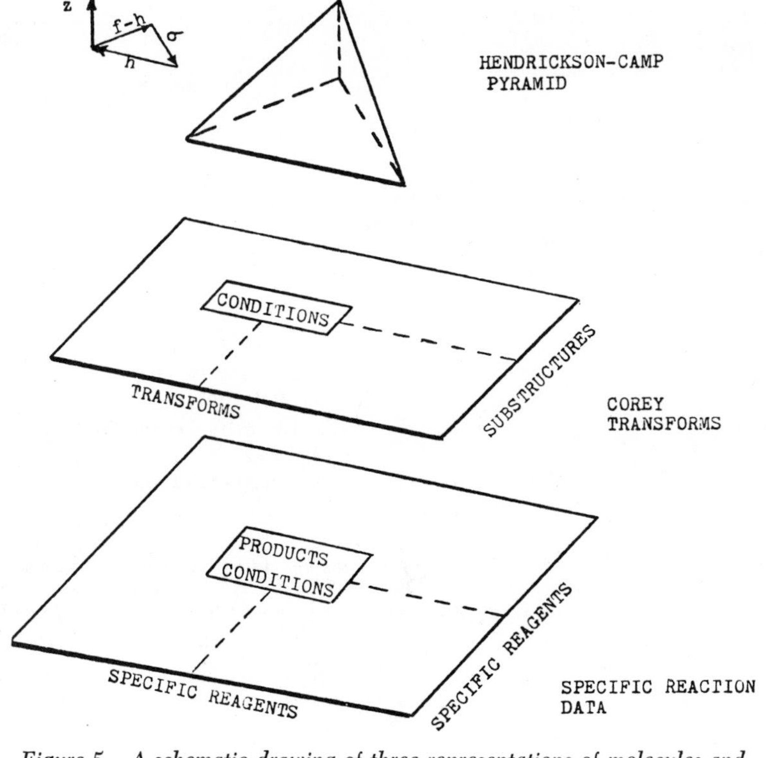

Figure 5. A schematic drawing of three representations of molecules and their reactions

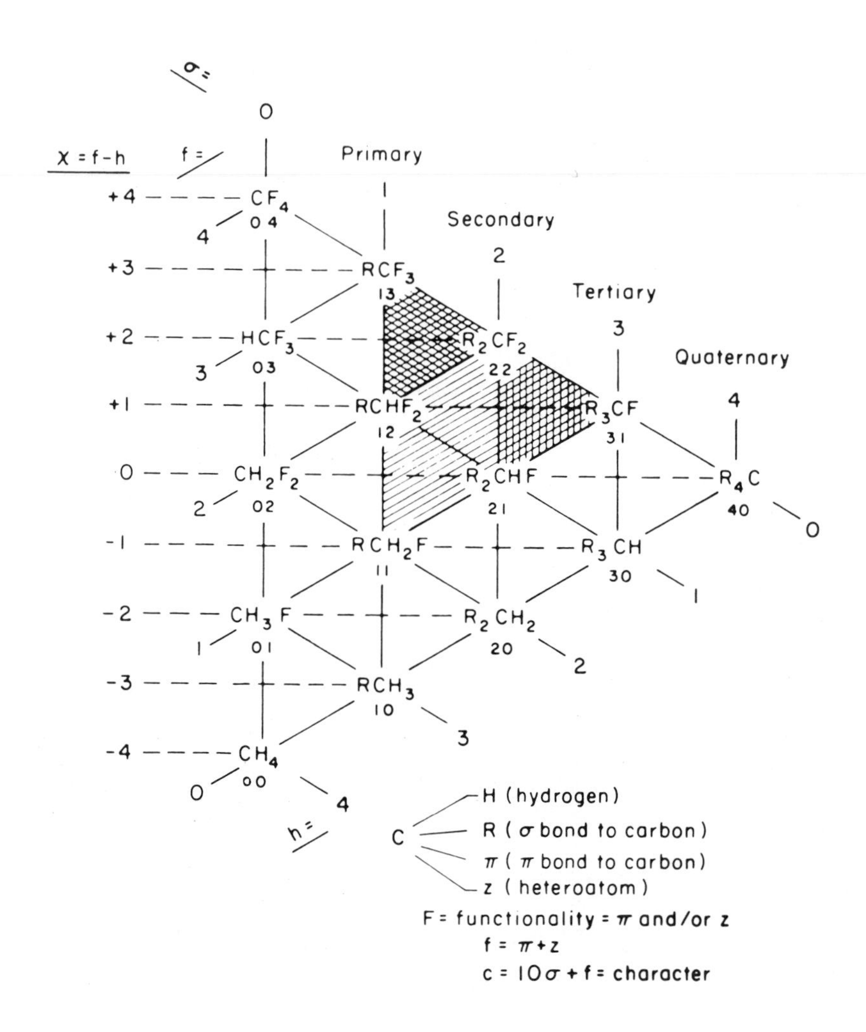

Figure 6. Character triangle for carbon sites and interconversions. The character C, is shown beneath each carbon site. The shaded areas indicate possible double bond sites (C = C). ▨ ; possible triple bond sites (C ≡ C), ▧ ; and possible aromatic sites ▥ .

into a pyramid. It is this representation that is shown in Figure 5.

The advantage of this representation is that it is very compact. Hendrickson's triangle has only 15 sites and 70 "reactions." Camp's pyramid has 24 sites and 132 reactions. The development of computer codes which utilize these representations is extremely simple. The representation is also able to generate "all" possible reactions which lead to a particular target molecule. The procedure could start at the target molecule and generate "all" possible precursors by using the reactions on the pyramid. Each site in the target is developed independently and certain dependencies must be checked (e.g. π bonds between atoms). In essence, this representation involves making (or breaking) every possible bond in the molecule subject to the valency and site constraints.

This generality is obtained with the loss of power to predict the consequences of each reaction step. The reactions are not classified by the exact nature of the functionality. Hence reactions with a functionality of f = 1 (alcohol, ether, olefin) are all treated similarly. In addition, some of the reactions are not known to have analogs in nature. These reactions just don't occur (or at least we haven't observed them yet).

Hendrickson's representation is an example of a generate-and-test problem-solving approach in which the generator is uninformed. It simply generates possible reactions and it is up to the tester (evaluator) to determine the feasibility and value of each step or path.

REPAS: A Case Study Using Hendrickson's Representation

Govind (4) developed a reaction path synthesis program based on Camp's generalization of Hendrickson's representation. The program was used to generate paths for a wide range of industrial molecules. For even a simple molecule such as acrylic (C=C-C(=O)-OH), well over 10,000 reaction paths were generated. The evaluation of these paths was a difficult task. Many of the paths included "strange" reaction steps for which no meaningful mechanism could be envisioned. While a few of these reaction steps provided interesting alternatives, there was a general lack of correspondence between these paths and those which experienced chemists and chemical engineers might execute in the laboratory, pilot plants, or industrial process. Our inability to be specific about each reaction step greatly limits the usefulness of this representation. In an abstract sense this approach could produce "all" reaction paths. However, the overwhelming number of combinations of functional groups and reactions severely limits the probability of finding a new and realizable reaction path. A representation of reaction is needed which is perhaps less general but has more specific chemical data and theory in it.

Substructure Representation

E.J. Corey and his co-workers (5) have developed a representation of molecules and reactions based on the combination of atoms in the molecule which change during known reactions. Molecules are represented by linked data lists. The lists contain the complete topological and atom type information for the molecule. Reactions are represented by transforms which indicate how substructures within a molecule change during a specific type of reaction. Classification of the reactions by only the atoms which change allows the transforms to be applied to a wide range of molecules which might contain these substructures. Figure 7 illustrates the general steps in applying a transform.

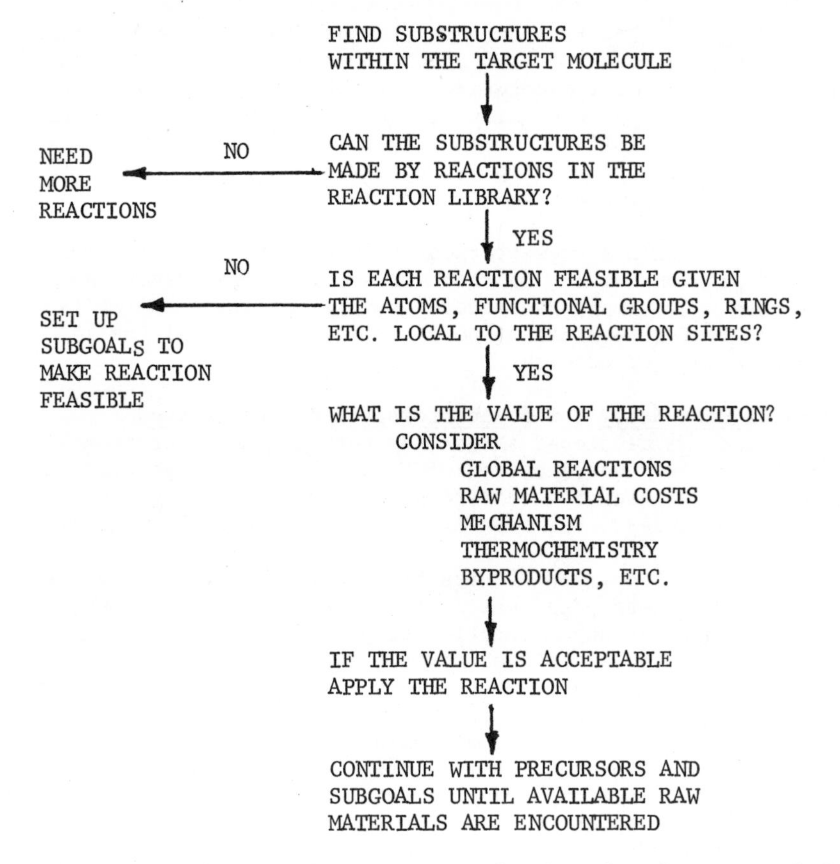

Figure 7. Steps in applying a transform based on the substructures which change during reaction

First, start with the target molecule and find all the substructures within the molecule for which transforms (reactions) are available. Then check to see if the transform is applicable for the specific molecule in question. In the transform there are three levels of evaluation. The first level is simply a check to ensure that the correct substructures are present. The second level of evaluation checks the atoms "local" to the reaction sites. If the local atoms prohibit the particular transform the reaction is either killed (i.e. no longer considered) or a subgoal is set up to change the atoms which are blocking the reaction. At the third level of evaluation the local atoms are checked to see how much they will improve or detract from the reaction. (e.g. if alpha is electron withdrawing add 10 to score.) In addition, it is important to check the global features of the molecule for competing reaction sites for the same transform, or for other reactions which might take place at the conditions required for the desired reaction.

Corey and co-workers (5) have developed a library of over 300 transforms of this type. Most of the secondary checks on the reactions (i.e. the affect of local attachments) are knowledgeable generalizations of the reviews published by Marsh (6), House (7), and Buehler and Pearson (8).

This approach has considerably more predictive power than the method of Hendrickson. Of course, it requires a larger data base. The generality of the approach is limited by the transforms in the data base and the secondary and tertiary checks made on each reaction. If the substructure transform is not in the data base it will not be included in any reaction path generated using this approach. If the secondary and tertiary checks are too conservative the reaction may not be used when it actually might work. It might be possible to systematically relax the local constraints (and use a method like Hendrickson's) in some future program. These reaction steps would then be the subject of a research and development program to see if they can be made to work.

Rote Memory

The most powerful predictor and least general approach is that of simply storing specific instances of known reactions along with the conditions under which the reactions were performed. The particular target molecule is then simply patterned matched with reactions in the data base.

Search Strategies

The main issue in search is whether the synthesis is planned by working backwards from the target or forwards from the starting materials. For many syntheses the target is a single molecule and the starting materials are many (commonly

10,000 or more). Hence working backwards is an efficient strategy. However, when the molecules are smaller, the set of starting materials may not be nearly so large. In petrochemical syntheses nearly all paths start with molecules such as ethylene, propylene, butane, etc., which often number less than 100. For cases like these in which the starting materials may be defined, a forward strategy may be more efficient. Powers and Jones (9) and Powers et al. (10) found that a forward-branching search strategy (discrete dynamic programming) was undoubtedly the best one for planning the chemical synthesis of bihelical DNA. The advantage of working forward from the raw materials is that the evaluation may be more accurately computed. When working backwards it is not possible to know the costs of precursors since their paths have not yet been developed.

In addition to search direction it is necessary to determine the order of enumeration of the synthesis tree. Depth, breadth and hybrid search procedures are all possible. For most current programs the evaluation is so uncertain that human interaction with the program during execution is necessary and undoubtedly more fruitful.

Applications to Industrial Reaction Paths

Most current programs have been aimed at laboratory syntheses. The detail and data required for industrial paths is not commonly in these programs. Many of these programs start with raw material molecules which are the targets of industrial chemists. We are intrigued by the possibility of applying these ideas to smaller molecules.

To our knowledge, only two applications of these techniques have been made to industrial reaction problems. The Chioda Chemical Engineering and Construction Company in Tokyo, Japan has used a program called CHEMONICS in which known specific reactions numbering approximately 100 are stored. The thermochemical properties, and in some cases kinetic parameters, are known for these reactions. The reactions are mixed and matched by the user with the program performing the analysis of the reaction network.

In the following sections we describe a prototype program, REACT, which is currently being tested on a number of industrial problems.

REACT -- A reaction path synthesis program for the petrochemical industry.

We are developing a program for use in the petrochemical industry. The features of the program are

1. Molecule Representation: Connection Matrix

For example,

READ TARGET MOLECULE	C=C=C=& (with O double-bonded above the third C)
COMPUTE CONNECTION TABLE	6 0 0 2 0 0 4 2 0 0 0 2 4 1 0 2 0 0 4 1 0 0 0 1 8
FIND FUNCTIONAL GROUPS AND POSITIONS	42 24 C=C

The current size limit is twenty carbon sites.

2. Reaction Representation
   - Substructure Transforms
     - Primary Evaluation on Substructure Presence
     - Secondary Evaluation on
       - Local Atoms, Rings, Functional Groups and Conditions.
     - Tertiary Evaluation on
       - Global Atoms, Rings, Functional Groups, Competing Sites for the Desired Transform, Impurity Reactions
   - Conditions Included
     - Solvent, Catalyst, Phase, Concentration(s), Temperature, Pressure, Time

   Currently 300 industrially important substructure transforms are in the data base.

3. Search Direction: Backwards from Single Target Molecule.

4. Search Strategies
   a. Depth-first guided by transform scoring function
   b. Breadth-first (exhaustive or with breadth control heuristics)
   c. User driven by interaction through a teletypewriter

5. Evaulation
   a. Feasibility by transform constraints
   b. Heuristic scoring at transform level
   c. Stoichiometry with raw material costs
   d. Side reactions
   e. User

6. Input/Output
   - Input: Free format input of the molecule in a 15 row x 15 column drawing box. Input by cards or teletypewriter
   - Output: Synthesis tree or reaction paths on line printer or teletypewriter

7. Program Features:
   FORTRAN, some linked list data structures, reactions are subroutines, interactive or batch, 2500 lines of code.

Example: ε-caprolactam

The development of new reaction paths to nylon monomer intermediates is an active industrial research area. Millions of kilograms of these monomers are produced each year and even small fractional improvements can have large economic impact. The uncertain future of feed stocks derived from crude oil has renewed interest in alternate reaction paths. In addition, the search for reaction paths with lower energy costs is intensifying. The following example illustrates the use of the REACT program to generate reaction paths which lead to ε-caprolactam. Over 500 paths were generated in less than 10 CPU minutes on an IBM-360/67 computer. Several of the paths are illustrated on Figure 8. This study was performed as part of an overall review of a chemical company's position in this monomer area. The results of the study indicated several new research directions which the research and development teams are now pursuing.

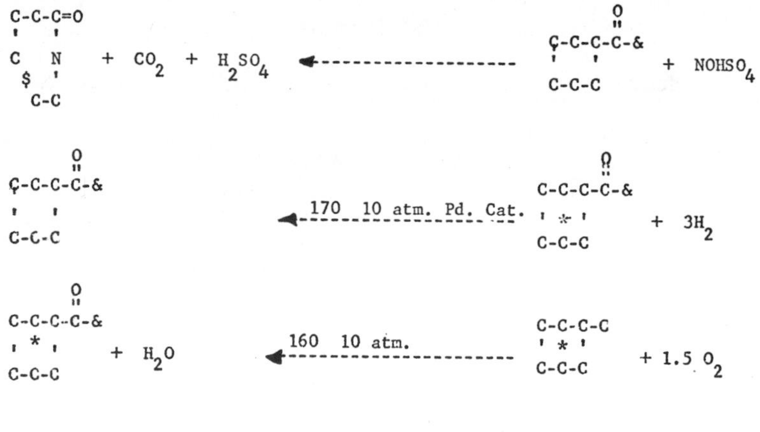

```
C-C-C=O                                                    O
'   '                                                      "
C   N   + CO2  + H2SO4  <-------------------          C-C-C-C-&
 $  '                                                 '   '        + NOHSO4
  C-C                                                 C-C-C

    O                                                      O
    "                                                      "
C-C-C-C-&                                             C-C-C-C-&
'   '              <--170  10 atm. Pd. Cat.--         ' * '        + 3H2
C-C-C                                                 C-C-C

    O
    "
C-C-C-C-&                                             C-C-C-C
' * '    + H2O     <--160  10 atm.-----------         ' * '
C-C-C                                                 C-C-C        + 1.5 O2
```

*Figure 8. Reaction path output from the REACT program: & = —OH, $ ≡ (sigma bond), * ≡ aromatic ring*

```
C-C-C=O                                                      C-C-C=O
'  '                                                         '  '
C  N    +  (NH4)2SO4         <-------------------------     C  N-H.H2SO4  + 2NH3
 $ '                                                         $ '
  C-C                                                         C-C


C-C-C=O
'  '                             H2SO4(SO3)                  C-C-C=N-&
C  N-H.H2SO4                 <-------------------------      '  '
 $ '                             Beckmann Rearr.             C-C-C
  C-C


C-C-C=N-&                                                    C-C-C=O
'  '     +  (NH4)2SO4 + H2O  <-------------------------      '  '   + NH2OH.H2SO4 + 2NH3
C-C-C                                                        C-C-C


C-C-C=O                          450  Zn. Cat.               C-C-C-&
'  '      +  H2              <-------------------------      '  '
C-C-C                                                        C-C-C


C-C-C-&                          150     Cat.                C-C-C-&
'  '                         <-------------------------      ' * '   +  H2
C-C-C                                                        C-C-C



  C-C-C=O
  '  '                           H2SO4                        C-C-C-N-&.HCl
  C  N     +    2HCl         <-------------------------       '  '
   $ '                           Beckmann Rearr.              C-C-C
    C-C


 C-C-C-N-&.HCl                   HCl  Light                   C-C-C
 '  '                        <-------------------------       '  '  +  NOCl
 C-C-C                                                        C-C-C


   NOCl  +  H2SO4            <-------------------------   HNOSO4 + HCl

  2HNOSO4 + H2O              <-------------------------   2H2SO4 + N2O3

   N2O3  +3H2O               <------------------------- 2NH3  +  3O2
```

Figure 8. Continued

Program Maintenance

The ability of programs like REACT to generate reasonable and potentially interesting reaction paths depends on the number and quality of reactions in the data base. A strong commitment to reaction documentation needs to be initiated within a company if the computer program is going to be useful. The speical reactions known to the company as well as reactions reported in the open literature need to be considered.

We have found that an intensive two-day training program is commonly sufficient to teach research and development chemists and chemical engineers how to code reactions for use in the program. A bimonthly review and generalization of each reaction should be performed by a group which includes the program specialists and chemists and chemical engineers drawn from across the company. Preliminary indications are that this approach is an effective way of actively capturing and applying a wide range of reaction know-how that is now sometimes lost in company reports or in the open literature.

More effort needs to be made in the area of systematic evaluation of reaction paths. These programs simply need to do more evaluation. To spend five minutes to generate 500 paths is simply not enough time. We need to see if it is possible to spend more time trying to find better paths. This need for systematic evaluation is going to push our understanding of approximate quantum mechanical calculations.

LITERATURE CITED

1. Rudd, D.F., G.J. Powers, and J.J. Sirrola, "Process Synthesis," Prentice-Hall, N.J., 1973.
2. Hendrickson, J.B., J. Am. Chem. Soc., (1971) 93, 6847.
3. Camp, D.F., MSChE Thesis, Massachusetts Institute of Technology, Cambridge, Massachusetts, 1974.
4. Govind, R., MSChE Thesis, Carnegie-Mellon University, Pittsburgh, Pennsylvania, 1976.
5. Corey, D.J., W. Todd Wipke, R.D. Cramer III, and W.J. Howe, Science, (1972) 166, 440.
6. Marsh, J., "Advanced Organic Chemistry: Reactions, Mechanisms, and Structure," McGraw Hill, New York, N.Y., 1965.
7. House, H., "Modern Synthetic Reactions," Benjamin, Menlo Park, California, 1972.
8. Buehler, C.A. and D.E. Pearson, "Survey of Organic Syntheses," Wiley-Interscience, New York, N.Y., 1970.
9. Powers, G.J. and R.L. Jones, AIChE J., (1973) 19, No. 6.
10. Powers, G.J., Russell Jones, George Randall, Marvin Caruthers, J. Van de Sande, and H.G. Khorana, JACS, (1975) 97, 875.

5

SECS—Simulation and Evaluation of Chemical Synthesis: Strategy and Planning

W. T. WIPKE, H. BRAUN, G. SMITH, F. CHOPLIN, and W. SIEBER
Board of Studies in Chemistry, University of California, Santa Cruz, Calif. 95064

1. INTRODUCTION

As the field of organic synthesis grows in sophistication it is appropriate that some thought should be given to the problem solving processes used in designing a synthesis and in selecting a synthesis from among many. Just as mathematicians have found the computer useful in studying the structure of finite groups and in searching for and testing new proofs,[1] so too the organic chemist is finding the computer useful in exploring the hypersurface of syntheses. Because actual execution of a synthesis is time-consuming and costly, the selection of which synthetic sequence to use becomes very important. The "Eureka Syndrome" leads one to stop generating alternative approaches after the first attractive solution is formulated,[2] but for an informed selection of the "best" route to try in the laboratory, one should examine several possible routes. Some of the advantages of computer-assisted design derive from the fact that the computer is not subject to the "Eureka Syndrome" or other human biases unless it is directed to be. A more complete discussion of these advantages is given elsewhere.[2]

Another important benefit of developing a program for computer-assisted design of synthesis is that in building such a program we have the opportunity to study the analytical processes chemists use in synthesis design and in some ways the computer program allows us to test our understanding of these processes and principles and the completeness of those principles. In this way research in computer-assisted design of syntheses can be expected to contribute to manual analysis through better understanding of principles and processes.

Our goal, given target T, is to generate good syntheses by working backward from the target toward simpler compounds (problem type III).[3] This paper is concerned mainly with this

I Start $\xrightarrow{?}$?

II Start $\xrightarrow{?}$ Target

III ? $\xrightarrow{?}$ Target

type III problem, as are the other papers in this symposium. However techniques developed for type III problems are also applicable to types I and II as we have previously demonstrated.[4]

Our goal state, the target molecule, is a specific organization of atomic nuclei and electrons. The target molecule is connected by chemical transformations to many other states which we call precursors. Some of the precursors are more complicated than the target (i.e., more difficult to synthesize than the target) and others are simpler than the target (see Fig. 1). The precursor and target are not required to have the same number of atoms of each type since there may be addition of a reagent to P or fragmentation of P to form T.[5] The physical laws of the universe

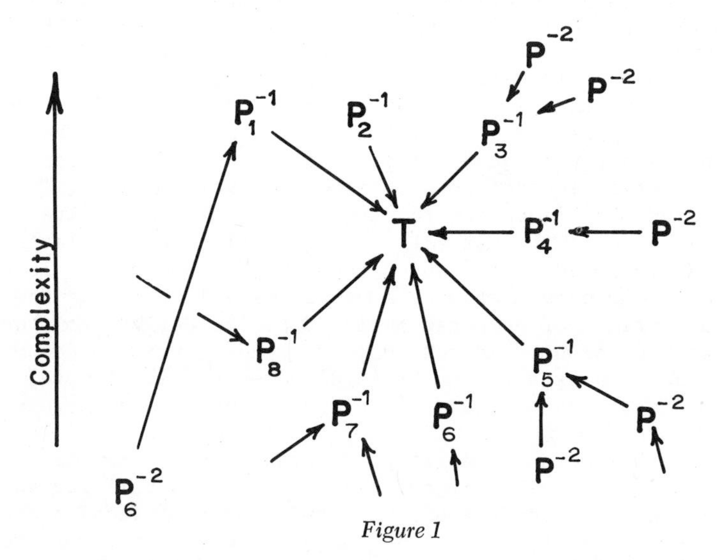

Figure 1

define all possible precursors which by some chemical change can lead to the target T, but we have little knowledge of this <u>absolute space</u> of states and most of what we have is gained by analogy and extrapolation from chemical transformations observed in related systems. Laboratory exploration of the <u>absolute space</u> only one step away from T would require subjecting <u>all compounds</u> to <u>all reaction conditions</u> and recording those which yield T as a product!

The space is very large as the example below illustrates.

$$\text{P} \xrightarrow[\text{2) } Me_2S]{\text{1) } O_3} \text{T} + \text{RCHO}$$

If cyclohexanone is the target, by arbitrarily varying R one can create as many precursors as one wishes, all of which do form cyclohexanone under these conditions, and hence are real states connected to T in the absolute synthesis space.

Now the purpose of the computer-assisted design program is to help the chemist visualize, explore, and evaluate this space. We would like to explore in the direction of simpler compounds (toward the bottom of figure 1) realizing that some increase in complexity of the precursors must be permitted because they may eventually lead to even simpler compounds, eg. P_1^{-1}, P_6^{-2}. The completeness of a synthesis program is the percentage of paths in absolute space which are generated (considered) by the program. The accuracy of the program is the percentage of paths generated by the program which are contained in the absolute space, since if a path does not belong to the absolute space it is an erroneous prediction.

Completeness assures no good solutions will be missed, and accuracy assures that predicted solutions are realistic and have high probability of working in the laboratory. In addition we generally want efficient solutions consisting of few steps and high overall yields. This paper describes how we approach these goals in the SECS program.

2. DESCRIPTION OF SECS SYSTEM

The SECS project for Simulation and Evaluation of Chemical Syntheses was initiated in 1969 to focus on stereochemistry, and the spatial and electronic aspects of chemistry, proximity, steric effects, and stereoelectronic effects, which the first synthesis program OCSS-LHASA[6,7] did not consider. Our goal was a program which would be useful in complex polyfunctional polycyclic synthesis, where success is dependent on selective reactions and selectivity is highly dependent on the steric and stereoelectronic nature of the reacting centers. SECS 1.0 was first demonstrated publically via teletype at a Gordon Conference July 1972, and SECS 1.5 was used with a DEC GT40 graphics terminal over the trans-Atlantic cable at the NATO Advanced Study Institute in Holland, June 1973.[3] Since 1975 a version of SECS has been available to the general public over TELENET and TYMNET. This paper describes the current version as of April 1976, SECS 2.4.

SECS 2.4 occupies 130K 32-bit words of memory, but when overlayed fits in 48K. Additionally the program uses about 1M bytes of auxillary disk storage for permanent and temporary files. SECS is written in FORTRAN IV and runs on PDP-10, PDP-20, UNIVAC 1108, IBM 370, and HONEYWELL-BULL systems.

To run the SECS program one needs only a teletype terminal or a DEC GT40 graphics terminal and access to a telephone. The chemist enters the target molecule on a GT40 system using the light pen (an integral part of the GT40 terminal). The structure is simply drawn with the light pen just as one would with a pencil. Details of the process has been given earlier.[3] SECS is carefully optimized to make such interactive graphical input efficient even over slow communication lines of 300 baud. For example, as each new bond is drawn in with the light pen, only the new bond is transmitted from the host computer rather than the entire updated molecular structure. For pictures of the input display the reader is directed to reference 3.

All functions of the SECS program can also be invoked from a teletype device or alphanumeric CRT terminal. Structural input of 1 from the teletype is illustrated by the dialog in Figure 2.

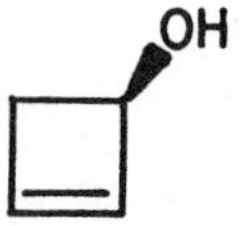

OH

1

When SECS begins the user is given an opportunity to see what changes have recently been made to this version. Then it asks if there is a RESTART file, this is to allow continuation from some previous session. If this is a new problem this is signaled by a carriage return (CR). The colon prompt indicates the SECS executive is executing. If the user types "HELP," SECS lists all commands available to the user at this point. TTYIN is then requested which asks for a name for the molecule and the number of non-hydrogen atoms. The user then enters the molecular connectivity by specifying a path of connected atoms. Note the multiple bond is indicated by giving the path between atoms 4 and 5 twice. All atoms are assumed to be carbon until specified otherwise. In the example, atom 1 was changed to oxygen. Finally the stereochemistry of the structural diagram is conveyed. Approximate atomic coordinates are entered to complete the "structural diagram." Z values may be entered if one wishes to fix the conformation, but normally the SECS model builder generates these when needed. The TPLOT command prints a

```
.RUN SECS

SECS VERSION 2.4, 4/1/76
LIST CHANGES (Y OR N)?

RESTART FILE NAME (OR CR):
:TTYIN
MOLECULE NAME (A20)   :TEST
ENTER NUMBER OF NEW ATOMS: 5
TO CREATE BONDS, TYPE BOND PATHS;END WITH BLANK LINE
BONDS: 1 2 3 4 5 2
BONDS: 4 5
BONDS                  [user types carriage return]
DO YOU WISH TO CHANGE ATOM TYPES, CHARGES, OR STEREO INFORMATION
(Y OR N):Y
TO CHANGE ATOM TYPES, TYPE ATOM # AND NEW TYPE
READY: 1 0
READY:
CHARGES:  TYPE ATOM # AND CHG(+, ,-)
READY:
TYPE CHIRAL ATOM #, CONNECTING ATOM #, AND DIRECTION (U OR D)
READY: 2 1 U
READY:
TYPE ATOM # + X,Y,Z COORDINATES
READY: 1 20 10
READY: 2 10 10
READY: 3 0 10
READY: 4 0 0
READY: 5 10 0
READY:
:TPLOT
C----C>>>0
!    !
!    !
!    !
!    !
C====C
```

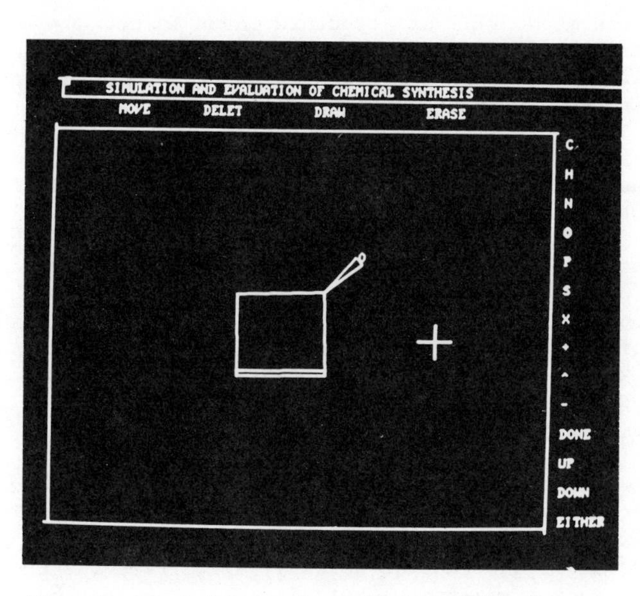

Figure 2. Input of 1 via teletype (top) vs. display (bottom)

structural diagram of the molecule on the TTY using the given relative coordinates. The ">" symbol indicates the OH is above the plane defined by the 4-membered ring.

Any errors in the structure can be easily corrected by the structure editor TTYED. TTYED can also be used to modify previously defined molecule files.

```
:TTYED
MOLECULE EDITOR - TYPE A COMMAND, OR HELP
ED:HELP
THE FOLLOWING ARE THE ONLY COMMANDS
RECOGNIZED BY TTYED

END - EXIT FROM TTYED
ADDED - ADD BONDS TO CURRENT STRUCT.
DELBD - DELETE BONDS
ATYPE - CHANGE ATOM TYPES
CHAGR - CHANGE ATOM CHARGES
STERE - CHANGE STEREO INFORMATION
XYZ - GIVE ATOM COORDINATES
ADDAT - ADD NEW ATOMS AND BONDS TO STRUCT.
DELAT - DELETE AN ATOM AND ALL BONDS TO IT
```

Figure 3. Molecule Editor Options listed by HELP command

This method TTY of structure input is not as elegant as that developed by Feldmann,[8] but it gives the user complete control over the layout or presentation of the structural diagram and includes stereochemistry.

The target molecule is now entered and ready for synthetic analysis. The molecule may be saved in a disk file with the WRITF command so the work of input will not be lost in the event of telephone disconnection. It can be read back in by a READF command. To begin analysis with the standard default strategies (which will be described later) the user simply types RUN or pushes light button "PROCESS." This generates the first level of the retrosynthetic tree. Each precursor is shown to the user on his terminal. He is then free to VIEW a precursor and PROCESS it further. Thus the standard usage of SECS is quite simple for the user.

With this overview let us now examine the components or modules of the SECS system shown in Figure 4. The executive is always present in memory together with one other module. This modular construction not only simplifies the program and makes it easy to maintain, but also allows us to keep memory requirements for execution of SECS below 48K 32 or 36 bit words.

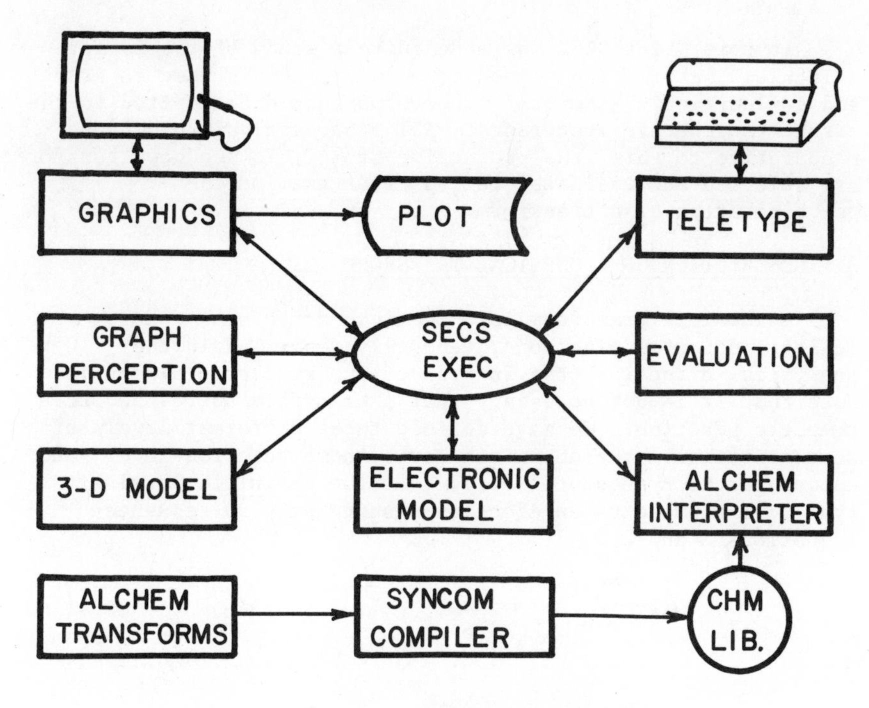

Figure 4. SECS 2.4 modules

After a structure has been entered into SECS, it is "perceived" for the presence of rings,[9] functional groups,[3,7] aromaticity, and is assigned a stereochemically unique name.[10] Molecular symmetry is also recognized at this time.[11] This is the limit of information that can be derived from only a structural diagram. A chemist would not stop at this point however, but would also derive various perceptions from the shape of the molecule, either from an implicit imagined 3-dimensional molecular model in the chemists mind or from an explicit physical Dreiding model which the chemist might build. Consequently SECS also has a model builder module[3,12] which constructs by constraint satisfaction techniques a 3-D model of minimum energy. SECS then has available Cartesian coordinates and Van der Waals radii of each atom, and thus "knows" the shape of the molecule. Now a second order perception determines the proximity of atoms, stereoelectronic orientation of bonds, strain energy, and even steric congestion.[13]

SECS also has an electronic model builder to perceive stabilization and locatization energies of conjugated systems as well as electron densities. This is discussed later in this paper.

At this point SECS has made quite a study of the target molecule. Information gained from this study is used to select and evaluate which chemical transforms should be applied to the target to generate precursors. All other analysis has been preparative to this, the productive stage. To see how transforms are selected and evaluated let us first examine the representation of a transform.

3. DESCRIPTION OF A CHEMICAL TRANSFORM.

A chemical transform is a chemical structural change, or redistribution of electrons, generally described in the analytical direction (the inverse of the synthetic direction. Note that it is not necessary that a transform correspond to a complete reaction. We have defined three different levels of representation: the *ab initio* level, name reaction level, and common reaction sequence level.[3] At the *ab initio* level a transform represents an electron-pushing step or sequence (equations 1 and 2).

$$A_1 - A_2 \Longrightarrow \overset{+}{A_1} + \overset{-}{A_2} \quad (1)$$

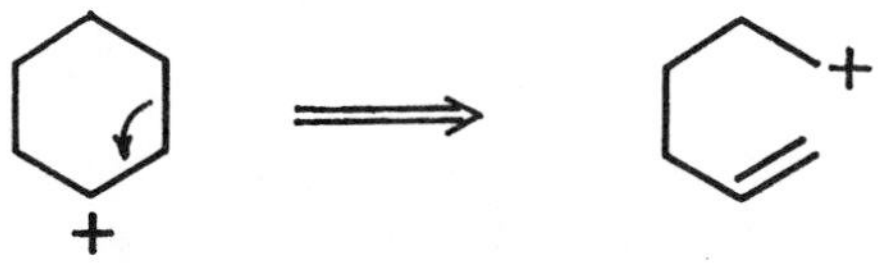

(2)

The left hand side represents a pattern which must exist in the target molecule and the right hand side, the pattern as it will exist in the precursor(s). The double shafted arrow means "implication," ie, seeing the left hand side in the target "implies" by this transform that it can be transformed to the right hand side.

The "name reaction" level, illustrated by the Aldol (equation 3), corresponds more closely with synthetic steps, and the precursors produced are normally stable isolable compounds. Of course a "name reaction" is a sequence of *ab initio* steps.

HO ⟹ O O (3)

$$RCOOH \Longrightarrow RCH_2COOH \quad (4)$$

Similarly a sequence of "name reactions" can be combined into a "common sequence" (equation 4).

Each level has its merits. The "ab initio" level is useful for discovering new reactions by searching for new sequences of electron pushing steps. The "name-reaction" level is most useful for novel total synthesis whereas the "reaction sequence" level is useful for rapidly generating classical syntheses, but has little innovative power.

3.1 TRANSFORM LEVEL AND COMPLETENESS. In order for a synthesis program to consider the complete space of problem solutions it is necessary to have a "complete" set of chemical transforms. If the "ab initio" level is used, then "completeness" is achieved with very few transforms like equation 1. The problem with the "ab initio" level is that the accuracy of predictions is quite low and very few paths in the synthesis tree generated are likely to succeed in the laboratory. One could improve the accuracy by adding to equation 1 constraints relating the feasibility of the process to the electronegativity and environment of atoms A_1 and A_2.

At the "name-reaction" level, thousands of reactions might be needed for reasonable completeness. The exact number is not known--we do not know how many "name reactions" there are since there is no agreed upon indexing system for reactions. The "name reaction" level has the disadvantage of requiring a larger knowledge base, but the advantage of providing the synthesis program with more accurate predictive capabilities. Except for mechanistic studies,[4] our synthetic work has been based primarily on the "name-reaction" level, because our interest is in generating accurate synthetic plans. By selecting this level we are saving the program the effort of rediscovering reactions that are already known, and we are preventing the generation of sequences of ab initio steps for which there is no known analogy. This is an example of tree reduction through greater chemical accuracy.

3.2 ALCHEM - A LANGUAGE FOR CHEMISTRY. In the SECS project we view a transform as a concise scientific statement of what a particular chemical transformation is and the factors which affect when the transformation will or will not occur. This packet of knowledge is structured through the language ALCHEM, an English-like language readable to chemist and computer. A transform contains the following types of information:

NAME
REFERENCE
SUBSTRUCTURE
PRIORITY
CHARACTER
CONDITIONS
SCOPE AND LIMITATIONS
MANIPULATIONS
"END"

The NAME is a text string for identification of the transform. REFERENCE is a leading literature reference substantiating the validity of the transform. The SUBSTRUCTURE describes a pattern which must be present in the target molecule for this transform to "fit." It describes the reaction site as it appears in the product, after completion of the synthetic reaction, generally including all required functional groups. For example, the substructure for the aldol reaction transform (eq. 3) could be "ALCHOHOL KETONE PATH 3", describing a pair of functional groups and a path of three atoms between them. This can alternatively be generalized using classes of functional groups: "ALCOHOL OXO PATH 3" where OXO = {KETONE, ALDEHYDE, ESTER}, or even more generally to: "DGROUP WGROUP PATH 3" where DGROUP = {ALCOHOL, AMINE, THIOL} and WGROUP = {KETONE, ALDEHYDE, ESTER, CYANO, NITRO, ACID HALIDE, etc.}. These examples involve a pair of groups in the target. Other reactions (e.g., reduction of a ketone to an alcohol) only involve one group in the target, hence the substructure only involves one group. The latter are called FGIs for functional group interchange. ALCHEM provides an alternative representation which allows description of any substructure, even where normal functional groups are not involved. A linear symbolic code has been developed for this purpose.[14] The aldol substructure could be represented as "O=C-C-C-OH/".

PRIORITY is a value representing the initial plausibility of the transform, before consideration of the reaction site context. Values range from -50 to 100. CHARACTER describes what kinds of structural modifications the transform can make, e.g., ALTERS GROUP, BREAKS RING.

CONDITIONS are generalized classes of reaction conditions, rather than specific reagents. The advantage being that any new reagent can always be represented as belonging to a general class. These are described later in this paper.

SCOPE and LIMITATIONS comprise the main body of the transform. They are "situation-action" rules which explore the context of the reaction site and alter the PRIORITY or other values to reflect the influence certain structural features have on the plausibility of this reaction. PRIORITY is one of eight vaiables to be altered in ALCHEM.

Figure 5 shows excerpts taken from the aldol transform to illustrate ALCHEM statements. Lines 2-5 represent the NAME, REFERENCE, SUBSTRUCTURE, PRIORITY, and CHARACTER, respectively. Group 1 is the DGROUP and Group 2 is the WGROUP. Atom 1 refers to the first atom on the path, the location of the DGROUP. Line 6 prevents the reaction if the DGROUP is an alcohol which is esterified. Line 7 says a nitrile is not a good electron withdrawing group in this reaction (WGROUP), and lowers the priority by 20. Line 8 evaluates the alkyl substitution on the attacking anion, lowering the priority for every substituent. Basic conditions are preferred (line 9), but if those conditions interfere with other sensitive groups in the structure, then acidic conditions can be used with a lowered priority (line 10). A manipulation statement (line 11) breaks bond 1 on the path connecting the DGROUP and WGROUP. Line 12 examines the hetero-atom in the DGROUP and calls it "(1)". If it is a quaternary nitrogen atom the transform is killed, but if it is a tertiary nitrogen a positive charge is placed on the nitrogen atom. This shows that manipulation statements can also be conditional as a result of a "IF...THEN" query. Line 13 puts in the multiple bond to the heteroatom. Line 14 increments the priority if there are no enolizable hydrogens adjacent to atom 1. Lines 15-18 lower the priority if the carbanion is required to attack the multiple bond from the most congested side. In evaluating congestion, SECS builds a 3-D model and integrates the accessibility of the reaction site.[15] Lines 15-18 also illustrate the powerful Set operations available in ALCHEM.[3] The word END signals the end of the transform. The aldol transform in the SECS transform library contains many more detailed scope and limitations statements.

The ALCHEM transforms are gathered together in unordered sequential source files. These source files are compiled by SYNCOM into efficient binary direct access CHM files complete with various directories.[16] The CHM files are later interpreted by SECS at run time. SYNCOM also provides very important syntax checking of the ALCHEM transforms. Since the transform library is not part of the SECS program, transforms can be added or modified without changing the SECS program. Since the transform library is not stored in main memory, the number and length of transforms in the library is effectively unlimited. The current library contains over 400 transforms.

A transform is not viewed as a program, but rather as a statement of facts. It does not contain GO TO's, LOOPs, SUBROUTINE calls, and does not reference any other transform. Each transform is independent of any other transform. Further a transform contains no strategies or heuristics about when its

```
 1)  ; HA-C-C-W ⇒ A=C H-C-W
 2)  ALDOL-COND
 3)  ;REF:  H.O. HOUSE, 'MOD. SYN. RXNS.', (1972) p 629.
 4)  DGROUP WGROUP PATH 3 PRIORITY 100
 5)  CHARACTER BREAKS CHAIN BREAKS RING
 6)       IF GROUP 1 IS ESTERX THEN KILL
 7)       IF GROUP 2 IS A NITRILE THEN SUBT 20
 8)       IF AN RGROUP IS ALPHA TO ATOM 2 THEN SUBT 10 FOR EACH
 9)       CONDITIONS BASIC AND NUCLEOPHILIC OR
10)       CONDITIONS ACIDIC AND NUCLEOPHILIC THEN SUBT 20
11)       BREAK BOND 1
12)       IF ATOM 1 IN GROUP 1 IS NITROGEN (1) THEN
             BEGIN IF (1) IS QUATERNARY THEN KILL
             ELSE IF (1) IS TERTIARY THEN ADD + TO (1)
             DONE
13)       MAKE BOND FROM ATOM 1 IN GROUP 1 TO ATOM 2 IN GROUP 1
14)       IF ATOMS ALPHA TO ATOM 1 OFFPATH ARE ALL ATTACHED
     &       TO 0 HYDROGENS THEN ADD 50
15)       IF ATOM 1 IS A STEROCENTER (1) THEN
16)          BEGIN IF ATOM IS ALPHA TO ATOM 1 OFFPATH (2)
17)          PUT (1) OR (2) INTO (1)
18)          IF ATOM 2 IS ON MOST HINDERED SIDE OF (1)
                THEN SUBT 50
             DONE
19)       END
```

Figure 5. Excerpts from the Aldol Transform

application will lead to a simplification of the synthetic problem. Strategies and transforms are separated completely, with the result that new transforms can be easily added without altering strategies, yet the strategies will use the new transforms when appropriate. This separation lends clarity and maintainability to the SECS knowledge base.[17]

ALCHEM is a powerful descriptive language. To date we have found no reaction that couldn't be represented. Part of the power of ALCHEM stems from the rich perceptual information base in SECS which is accessed by terms such as STERIC HINDRANCE, and the capability of SECS to evaluate these expressions in terms of models. In following sections we show how these expressions and models are used to achieve greater chemical accuracy and thus effectively reduce the size of the synthesis tree generated.

4. SYNTHESIS TREE REDUCTION THROUGH GREATER CHEMICAL ACCURACY

4.1 STERIC EFFECTS. A chemist predicts steric effects by evaluating the possible approach of a reagent to a three-dimensional model of the reacting molecule. SECS does the same thing. The SYMIN module constructs a 3-D space-filling model according to the required stereochemistry using molecular mechanics.[12] Any transform can invoke a steric calculation as illustrated in line 18 of figure 5. Of particular importance are reactions which create new stereocenters by attack on a double bond. For example, in the Uclaf synthesis of andrenosterone, implication of the ketone precursor from the β- alcohol is valid because it also implies hydride attack from the less congested side of the carbonyl which is preferred.

We developed a calculational model for this type of reaction to permit evaluation of steric accessibility and its inverse, congestion, on any type of molecular structure (see fig. 6).[15]

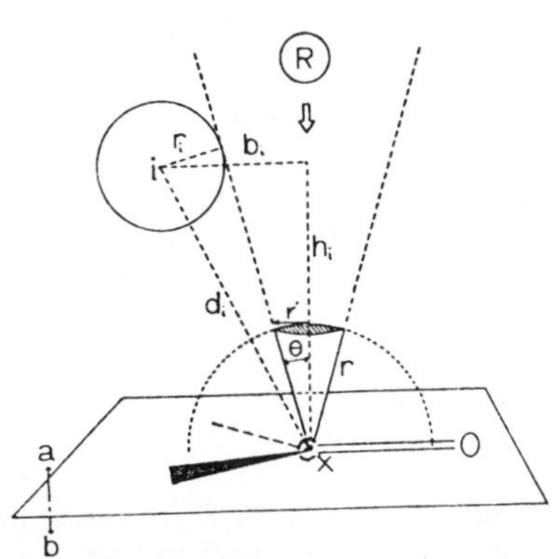

Figure 6. Cone of preferred approach of R to x allowed by atom i. The accessibility of x on side a with respect to i is defined by this solid angle and numerically equals the area on a unit sphere cut by this cone (shaded area).

Using this model to predict the major product from ketone reductions, we found predictions correlate well with experiment ($n = 34$, $r^2 = 0.924$). We also found this model applicable to epoxidation and hydroboration of olefins.[24] The steric ALCHEM statement (line 18, fig. 5) is evaluated via the steric calculation to determine if the reaction would create the required stereoisomer, and depending on the truth value, the appropriate conditional phrase is interpreted. Note this is comparison of congestion between the two sides of a double bond.

Because the calculated congestion is absolute, one can also make comparisons between functional groups within a structure. For example, in Corey's longifolene synthesis it was necessary to remove one ketone in the presence of another as shown below.[25] Normally this would require protecting the ketone which is to remain. Both ketones are saturated, but they differ in steric environment. The total congestions calculated by SECS indicate that the ketone which is to remain is already protected by the

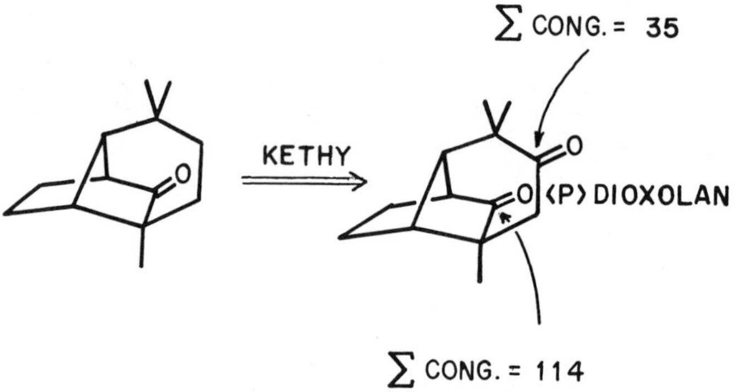

(4a)

greater steric congestion present, hence reactions should occur on the α,α-dimethyl ketone selectively and this was observed experimentally. This is expressed in ALCHEM by "IF STERIC AT GROUP 1 BETTER THAN KETONE ANYWHERE THEN..."

This approach of evaluating steric effects has proven effective[15] on a wide variety of molecular skeletons and is not restricted to any library of special ring systems. We believe SECS's predictive capability in an unknown system compares favorably with the predictive capability of most chemists.

4.2 ELECTRONIC EFFECTS. The electronic properties of conjugated systems could be stored in a library according to ring system and heteroatom substitution. Such an approach however fails when faced with a skeleton not contained in the library. For directing effects on benzene, one could use the Hammett equation[18] or topological rules,[19] but these cannot easily be extrapolated to polycyclic or heterocyclic aromatic systems. As a more general and flexible approach, we decided SECS should construct its own molecular orbital (MO) model of the conjugated system from which it could derive whatever properties it needed. The Hückel MO method was selected to minimize computational effort. Wtihin this framework, localization energies[20] have proved more successful than electron densities[21] in predicting aromatic substitution reactions.

The HAREHM module recognizes substituents on aromatic rings and heteroatoms in the rings, and assigns to each of them h-parameters characterizing their electron withdrawing or donating properties.[22] For each aromatic ring system HAREHM analyzes the possible substitution reactions: for each substituent on the ring system, the substituent is removed and the electronic energy is calculated. Then for each free position on this modified structure the free atom is temporarily removed, breaking up the conjugated system, and the energy again calculated. The difference between these energies is the localization for attack at that free position. This is computed for anion, radical, and cationic intermediates when appropriate, and for all free positions. If the position having the lowest localization energy is the same position as the original substituent X then it is correct to infer that kinetic attack of X will occur in the correct position. See examples in equations 9 and 10. HAREHM can treat structures with multiple conjugated systems.

Within an aromatic substitution transform the terms RLENERGY, NLENERGY, and ELENERGY refer to the radical, nucleophilic, and electrophilic localization energies respectively. Several types of statements use these terms, e.g.:

```
IF ELENERGY ON ATOM 1 BETTER THAN ATOM 2 ELSE KILL                    (5)

IF ELENERGY ON ATOM 4 BETTER THAN (1) THEN ADD 20                     (6)

IF NLENERGY ON ATOM 4 BETTER THAN (1) THEN                            (7)
     BEGIN ADD 40
     CONDITIONS NUCLEOPHILIC AND BASIC
     DONE
     ELSE BEGIN IF ELENERGY ON ATOM 4 BETTER THAN (1)  THEN
          BEGIN ADD 20
          CONDITIONS ACIDIC AND ELECTROPHILIC
          DONE
     DONE

IF ELENERGY ON ATOM 4 BETTER THAN (1) THEN CORRECT                    (8)
```

Statement 5 shows comparison of electrophilic localization energy (LE) on 2 atoms, whereas in statement 6 the LE of atom 4 is compared to the LE's of each atom in the set of atoms (1). The compound statement 7 selects nucleophilic conditions if possible, otherwise electrophilic conditions. In statement 8, "CORRECT" adjusts the priority upward or downward according to whether the LE is lower or higher than the LE of benzene. The magnitude of change is proportional to the difference between the two LE's. Factors such as steric effects on thermodynamics are described by other statements in the transforms. The examples below illustrate control by electronic effects. The numbers in equation 10 show calculated LE values.

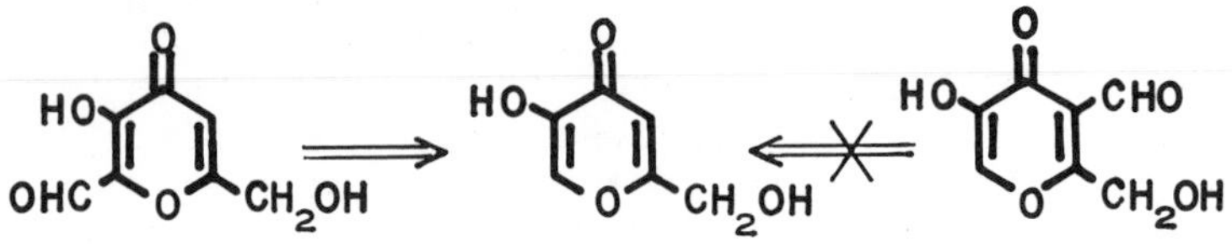

(9)

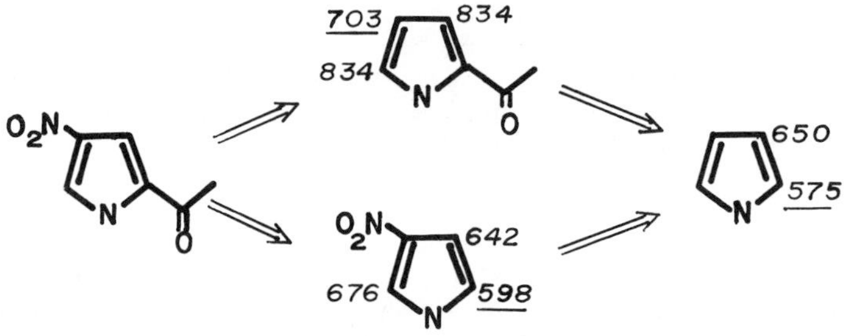

(10)

The accuracy of the HMO method decreases as the number of heteroatoms in the ring system increases, but the same can be said for predictions by chemists in complex cases. The HMO method has satisfied our need for a rapid, flexible, general approach.

4.3 RECOGNITION OF INTERFERING FUNCTIONALITY AND PROTECTING GROUPS. An important part of evaluating the applicability of a given transform is determining whether the conditions required for the transform will also cause unwanted reactions elsewhere in the molecule. This is in fact an important selection rule which helps further reduce the potential solution space and prevent some possible orderings of transforms. However, it is too harsh simply to kill any transform in which there is an interference when that interference could be avoided by modifying the group. We have also found that it is desirable for the user to be notified of such interferences, because he may not know the exact conditions and can not evaluate interferences easily simply looking at the precursor, given the name of a transform.

Thus, in 1973 we incorporated protecting groups into SECS 2.0, increased the number of functional groups recognized, and differentiated between functional group environments. (See Table I for a listing of the sensitivities of differentiated functional groups to different reaction conditions.) Each functional group is classified into exactly one category which best describes it. The definition of reaction conditions was also changed in SECS to cover more types of conditions. There are many ways to describe reaction conditions, one can describe physical parameters such as temperature, time, pressure, acidity, redox potential, polarity, etc., and let a given reaction be represented by a sum of such parameters; or one can define reagent prototypes; or one can define equivalences between reagent prototypes and physical parameters. The current categories in SECS 2.4 are shown in Table II. Each category is subdivided into slight, normal, and strong, corresponding to the continuous physical properties (represented in Table I by -, = and * respectively), but for some chemical categories the subdivisions represent different reaction types. A group sensitive to slightly acidic conditions is usually more sensitive to strongly acidic conditions, but a group sensitive to slightly halogenating conditions may be stable to strongly halogenating conditions because the two halogenations are mechanistically completely different. All reaction conditions used in a reaction up to the next workup (one pot) are specified in one CONDITION statement in ALCHEM. Thus if there is more than one step (workup) in a transform, there would be separate condition statements for each set of simultaneous conditions. The reason is that a functional group might be stable to every single reaction step, but not to an attack of all conditions at the same time. The current version takes the condition categories as independent and additive, so one specifies all the pure physical properties (pH, temp., solvent) and only the relevant chemical properties.

The basic chemical information used for functional groups, protective groups, sensitivities and reaction conditions is represented by binary sets, and manipulated with logical set operations, AND, OR, XOR, and SUBSET.[3] A binary set containing the current functional groups is established during perception of the target:

GROUPS 1001000000 (group types)

Every bit position corresponds to one of the categories of functional group, 1 means present, 0 means absent. Every transform contains a condition statement. In the transform corresponding to the synthetic reduction of a ketone shown below, the conditions are alternatives from which only one set of conditions is

Table I Sensitivities of Functional Groups to Reaction Conditions

		BAS	ACI	OXI	RED	HOT	OMT	PHO	SOL	HYD	HAL	NUC	ELE
1	CARBOXYLIC ACID				*		-=*					*	=*
2	DECARBOXYLATIVE ACID		=*		*	=*	-=*					*	=*
3	ACID ANHYDRIDE	-=*	-=*		=*		-=*		*			=*	
4	ACID HALIDE	-=*	-=*		=*		-=*		*	-=*		=*	
5	PHENOL			-=*			-=*				=*		=*
6	ENOL				*		-=*				*		=*
7	BENZYL,ALLYL ALCOHO		=*	-=*	=*	*	-=*			=*			=*
8	PRIM ALCOHOL		*	-=*			-=*						=*
9	SEC ALCOHOL	*	*	-=*			-=*						=*
10	TERT ALCOHOL	=*	=*				-=*						=*
11	AROM ALDEHYDE			*	=*		-=*	*		=*		=	*
12	A,B UNSAT ALDEHYDE			*	=*		-=*	*		=*		=	*
13	CH 3,R CH2 ALDEHYDE	=*	*	*	=*		-=*	*		=*		=	*
14	R2 CH2 ALDEHYDE	*		*	=*		-=*	*		=*		=	*
15	R3C ALDEHYDE			*	=*		=*			=*		=	*
16	PRIM AMIDE	=*	*		*		=*					*	=*
17	SEC AMIDE	*	*		*		=*					*	=*
18	TERT AMIDE				*		*					*	=*
19	AROM AMINE			-=*			-=*				=*		
20	ENAMINE		=*	*	=		-=*			=*	=*		
21	BENZYL,ALLYL AMINE	*	*	=*	*	*	-=*			=*			
22	PRIM AMINE			=*			-=*						-=*
23	SEC AMINE			=*			-=*						-=*
24	TERT AMINE			*									
25	C=C AROMATIC							*					
26	C=C CONJ TO C=C	*	*	*				=*		=*	=*		*
27	C=C CONJ TO W-GROUP	*	*	*			=	=*		=*	=*	-=	
28	C=C ISOLATED	*	*	*						=*	=*		*
29	CH3,RCH2 ESTER	=*	=*		*		=*					*	*

		BAS	ACI	OXI	RED	HOT	OMT	PHO	SOL	HYD	HAL	NUC	ELE
30	R2 CH ESTER	*	*		*		=*					*	*
31	OTHER ESTER				*		=*					*	*
32	A-HALOGEN ETHER	*	*		=*	*	=*			=*		*	
33	PHENOL ETHER	*	*				=*						
34	ENOL ETHER					*	*			=*			
35	BENZYL, ALLYL ETHER		*	*	*	=*	*			=*			
36	ALKYL ETHER												
37	ARYL HALIDE						=*	*		=*			
38	VINYL HALIDE						=*						
39	BENZYL, ALLYL HALIDE	-=*	-=*		*	*	-=*			=*		-=*	
40	PRIM HALIDE	*	*				=*					-=*	
41	SEC HALIDE	*	*				=*					-=*	
42	TERI HALIDE	-=*	-=*				=*			*		-=*	
43	ARYL KETONE				=*		=*			=*		=	*
44	A,B UNSAT KETONE				=*		*	=*		*		=	*
45	CH3,RCH2 KETONE	*	*		=*		-=*	*		*	*	=	*
46	R2 CH KETONE	*	*		=*		=*	*		*	*	=	*
47	R3C KETONE				*		*						*
48	QUINONE				-=*	*	=*	=*		=*		=*	
49	NITRO				-=*			*		-=*			
50	NITRILLE	=*	=*		-=*		=*			*		*	
51	ALKYNE-H		*	*		*	-=*	*		-=*	=*		*
52	ALKYNE-R		*	*		*		*		-=*	=*		*
53	EPOXIDE	=*	=*		*		=*			=*	*	=*	
54	CYCLOPROPANE	*	*	*	*	*				*	*		
55	PHENOL,ENOL ESTER-X	-=*	-=*			*						=*	*
56	BENZYL,ALLYL ESTER-X	-=*	-=*			=*				=*		=*	*
57	TERT ESTER-X	*	*										
58	SEC,PRIM ESTER-X	=*	=*									*	
59	IMINE	=*	=*	=*	*	*	=*	*		=*	=*	=*	
60	IMINE-Z												

Table II

Definitions of

Reaction Conditions

STRONGLY	pH(pKa)>18
BASIC	12 - 18
SLIGHTLY	8 - 12
SLIGHTLY	2 - 6
ACIDIC	-2 - +2
STRONGLY	< -2
STRONGLY	$KMNO_4$, CRO_3
OXIDIZING	JONES REAGENT
SLIGHTLY	MNO_2, MILD SELECTIVE
SLIGHTLY	ZN/H+, MILD SELECTIVE
REDUCING	$NABH_4$, ALKYL HYDRIDES
STRONGLY	$LIALH_4$
STRONGLY	< -100
COLD	-100 - -20
SLIGHTLY	-19 - +10
SLIGHTLY	30 - 100
HOT	101 - 200
STRONGLY	> 200
SLIGHTLY	ORGANOCADMIUM AND ZINC
ORGANOMETALLIC	ORGANOMAGNESIUM
STRONGLY	ORGANOLITHIUM
SLIGHTLY	UNSENSITIZED 350 - 700 NM
PHOTOCHEMICAL	UNSENSITIZED 250 - 700 NM
STRONGLY	SENSITIZED
PROTIC	
APROTIC	
SLIGHTLY	DEACTIVATED CATALYSTS
HYDROGENATING	MILD CATALYSTS
STRONGLY	STRONG CATALYSTS OR/AND HIGH PRESSURE
SLIGHTLY	N-BROMOSUCCINIMID IN CCL_4
HALOGENATING	HALOGEN WITHOUT CATALYST OR R-X/$SNCL_4$
STRONGLY	HALOGEN WITH ACID OR BASE CATALYSIS
SLIGHTLY	SUBSTITUTION AT SATURATED CARBON AND MICHAEL
NUCLEOPHILIC	ADDITION AT CARBONYLS
STRONGLY	SUBSTITUTION AT CARBOXYLS (ADDITION-ELIMINATION)
SLIGHTLY	ALKYLATION WITH R-X
ELECTROPHILIC	ACYLATION WITH RCOX, $(RCO)_2O$, TSCL
STRONGLY	LEWIS ACID CATALYZED ALKYLATION OR ACYLATION

selected.

```
    ; CH2 => C=O
    ; WOLFF-KISHNER, THIOKETAL DESULFURIZATION, OR CLEMMENSON
    ;   REDUCTION OF A KETONE
    KETHYDROGENOLYSIS
    KETONE PATH 1 PRIORITY 50
    CHARACTER INTRODUCES GROUP
        IF ATOM 1 IS NOTE ATTACHED TO 2 HYDROGENS THEN KILL
        IF HETEROATOM IS ALPHA TO ATOM 1 THEN KILL
        IF ATOM 1 IS ALLYLIC THEN SUBT 70
        CONDITIONS STRONGLY BASIC AND REDUCING OR
        CONDITIONS ACIDIC AND REDUCING OR
        CONDITIONS SLIGHTLY HYDROGENATING AND NUCLEOPHILIC
        ADD O OF ORDER 2 TO ATOM 1
        END
```

CONDIT compares the set of conditions for the reaction (or reaction step) with the set of sensitivities of each group to those conditions, leading to a set of interfering groups (IGRPS) (acid interferes):

SENS (acid)	1000100010	(condition types)
SENS (ketone)	0010001000	
COND	0001100000	(condition types)
GROUPS	1001000001	(group types)
IGRPS	0001000000	

The program now looks for protective groups based on the following information:

NAME	2-OXAZOLINE	
PROGR	0001100000	(group types)
STABIL	1011010001	(condition types)
INTRO	0000000001	(condition types)
REMOV	0100000000	(condition types)
DEDUC	35	(priority change)

The group-set PROGR indicates the group types that can be protected by this particular protective group, STABIL contains the reaction conditions to which the protective group is stable. INTRO and REMOV contain the reaction conditions needed for introduction and removal of the protective group. DEDUC is a measure in priority value units of the expenditure needed for an application of that certain protective group, based on the number of steps, losses in yield, and difficulty in experimental procedure. The algorithm for choosing the proper protective group is based on the following criteria:

a) the PG must fit the FG
b) the PG must be stable to reaction conditions (all steps)
c) the PG should simultaneously protect as many as possible of the interfering groups
d) the PG should protect none or as few as possible of the non-interfering groups
e) the priority loss should be minimized

If there remain any unprotected interfering groups, an additional penalty is imposed which may kill the transform if the priority cutoff is high enough. If the transform was successful in all parts, the names of the used protective groups are stored with the manipulation instructions for the generation of the precursor from the current target. The atoms bearing a protective group are marked in the connection table. When the precursor is displayed a (P) appears near the starting atom of the functional group along with the name of the specific protecting group (the name can be suppressed by the user). (See equation 4a.) The protecting groups are automatically removed from the precursor when it is processed as the current target. Thus, the user is notified graphically of the fact that in this given transform certain groups must be protected, but the decision of when to put them on or take them off must be done after the entire sequence has been generated. At that time one would possibly select a different group which satisfies the needs of several transforms and minimizes operations. Work is under way to optimize protection for sequences of reactions and to generate subgoals for functional group interconversion for particularly troublesome groups, using principles of latent and equivalent groups.

Mutually reactive functional group combinations like acid halide and alcohol in the same precursor are recognized in the precursor evaluation routine. The user specifies to EVAL whether such precursors should be deleted.

5. TREE REDUCTION THROUGH SYMMETRY

In the absence of symmetry information, application of a transform to a structure possessing several elements of symmetry produces redundant precursors. The precursors must be created, the redundancies recognized by graph-matching or canonical naming, and finally the duplicate precursors must be deleted. With symmetry information, it is possible to avoid generating such redundant precursors. SECS perceives the complete stereochemical graph isomorphism groups of the target structure and uses this symmetry group in applying a transform so that the transform is applied in all possible unique ways without redundancy.[11] The synthesis of asterane[25] (figure 7) illustrates how recognition of molecular symmetry reduces the number of pathways generated.

Figure 7. Redundancy from molecular symmetry

In addition to utilizing molecular symmetry, it is also necessary to utilize transform symmetry where it exists. A transform possesses a symmetry element if the SUBSTRUCTURE possesses the element and it is preserved by the MANIPULATION and SCOPE/LIMITATIONS statements, e.g., the Diels Alder transform:

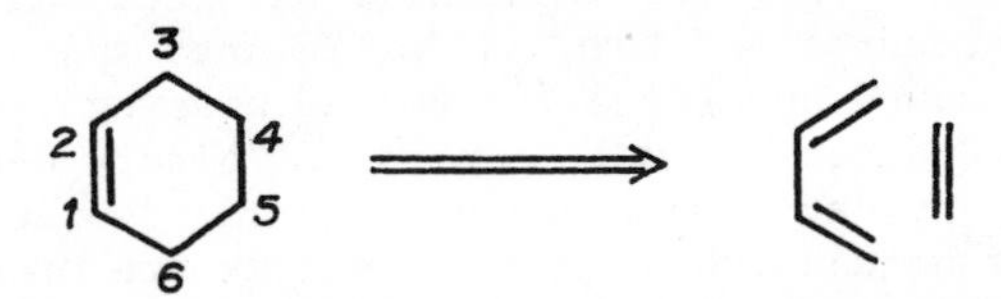

This symmetry is represented in the substructure, "C=C-C-C-C-C-@1<2,1,6,5,4,3>/", by the operator enclosed within angle brackets and prevents SECS from flipping the SUBSTRUCTURE over on the target and applying the transform again, which would produce an identical precursor.

SECS also uses the stereochemical graph isomorphism group within a transform to correctly determine if atoms or groups are chemically equivalent: "IF ATOM 1 AND ATOM 2 ARE EQUIVALENT THEN..." Complete details of the symmetry recognition algorithm and its application are given elsewhere.[11]

6. TREE REDUCTION THROUGH PRUNING

Precursors are screened independently by the EVAL module to eliminate illegal structures which might have been created by a faulty transform, and structures which are unstable or otherwise undesireable. The evaluation criteria listed below can be enabled or disabled by the user through the EVAL command.

1. Bredt rule violation
2. Antiaromatic ring
3. Cumulene
4. Triple bond in small ring
5. Trans olefin in small ring

6. Trans bridged ring system
7. Trans fused 3-membered ring
8. Valency violation
9. Di-ion of same charge
10. Four-membered ring
11. Unstable group combinations

The previous sections dealt with reducing the size of synthesis tree generated by increasing the accuracy of predictions, i.e., by making transforms more selective in their output. We now turn our discussion to how the program selects which transforms to apply.

7. CONTROLLING TREE GROWTH THROUGH STRATEGIES AND PLANS FOR TRANSFORM SELECTION

The objective of the SECS program is to find sequences of transforms which not only are chemically accurate, but also lead to "simpler" structures and form an "efficient" synthesis. We have for many reasons specifically avoided biasing SECS with preprogrammed sequences or their equivalent binary decision trees. Such binary decision trees produce rigid orderings of transforms, must be updated as new transforms are discovered, and are necessarily incomplete since they are manually programmed. Instead, our approach is to have SECS dynamically generate its own synthetic sequences guided by its current strategies using its current library of transforms. Strategies are high level principles of molecular construction and modification, and are independent of the transform library. When a strategy is applied to a target structure, goals are generated which describe the structural changes necessary to implement the strategy. Goals are also independent of the transform library. The null strategy generates no goals, hence selects no transforms.

7.1 THE OPPORTUNISTIC STRATEGY is the opposite of the null strategy, for it selects every transform which "fits" the target, i.e., selection by applicability, regardless of whether the precursor is simpler than the target. This generates all "legal moves" in chess terminology. A major problem with this strategy is that the large class of functional group interchange (FGI) and function group introduction (INTRO) transforms, are nearly always applicable and are thus frequently applied, but do not normally lead to simpler precursors. FGI's are the equivalent of the substitution operators in GPS.[27]

7.2 GOAL-DIRECTED SELECTION of transforms operates quite differently--the first consideration becomes "does this transform achieve one of my goals?", i.e., if it were applied would it be a move in the desired direction? Goals define the "desired direction." A goal is the difference between the current target structure and the desired structure or desired substructure. Transforms are then first selected as to whether they are relevant to the goal(s), and later are examined for applicability. Goals provide a sense of direction and a justification for expending computational resources. This enables a selective exploration of the space of syntheses, and promises a higher proportion of efficient syntheses among the solutions generated, the tradeoff being loss of completeness and possible loss of solutions. However, with good goals this loss is negligible. As the number of chemical transforms available to the program increases, the benefits of goal-directed selection of transforms become even more obvious.

7.3 GOALS ARE CREATED BY STRATEGIES. An important objective of the SECS project is to determine what constitute good strategies for synthesis. Many strategies are self evident but others are yet to be discovered. Our fundamental aim is to obtain simple precursors. Network oriented strategies focus on simplification of the molecular skeleton, in particular the ring systems. Fused ring systems are considered simpler than bridged or spiro systems owing to their greater ease of preparation. A good synthetic strategy is therefore to create the complex ring system from a simple fused system. Translated into the backward direction, this means try to break those bonds which lead to fused or acyclic ring systems with a minimum of appendages. The first statement of this strategy appeared in Corey's longifolene synthesis.[25] Thus bonds 1-7, and 3-4 are strategic, but so is

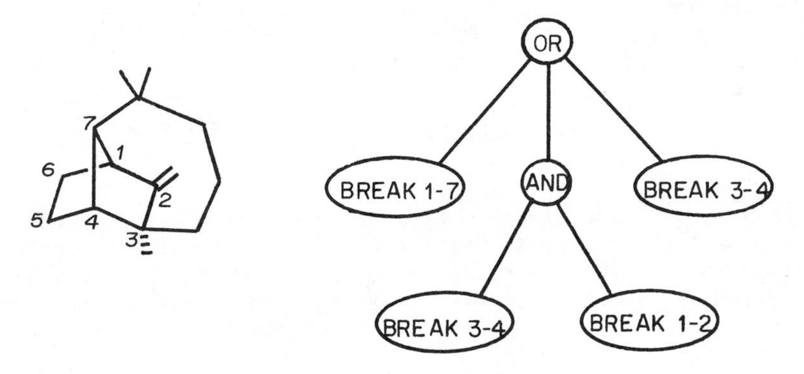

the pair of bonds {3-4,1-2}. In place of a set of strategic bonds used by other programs,[7] SECS represents this strategy as a logically structured goal list shown above.

Other strategies focus on simplifying the skeleton by reconnecting appendages to form a new ring or by joining opposite sides of large rings to form common rings. These strategies produce goals such as "MAKE 1-6". Note this goal can not be represented as a set of bonds since the bond 1-6 does not yet exist.

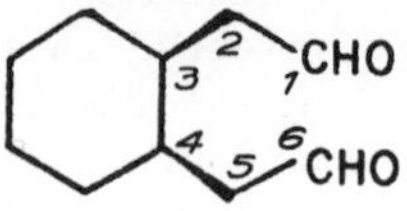

Symmetry based strategies are indicated in the syntheses of β-carotene which constructed bonds a and b,[28] c and d,[29] and e and f,[30] respectively. This strategy would produce the goal

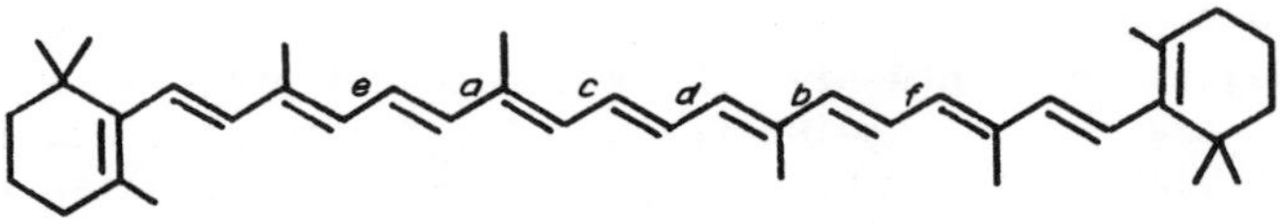

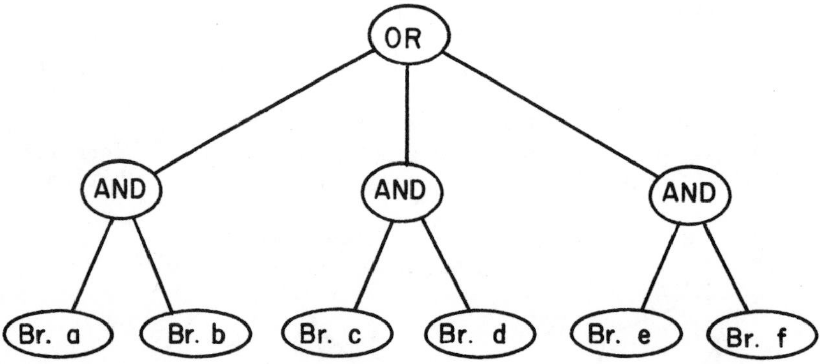

list shown. These examples illustrate how strategies create goals and why the goals are structured. There are of course many varied synthetic strategies, relating to many structural attributes. The goal list is the common denominator that relates these strategies to selection of transforms.

7.4 EXPLICIT GOAL REPRESENTATION. In SECS, goals are generated dynamically and represented explicitly, in contrast to being represented implicitly by a static programmed procedure. Explicit representation of goals has the advantages that goals can be rearranged in priorities, the order of goal satisfaction does not need to be the same as the order in which the goals were created, and the chemist can see what the goals are and can modify them interactively.

The goal list is a general list structure containing currently four types of entities:

1. Logical connectives
2. Functional group modifications
3. Structural modifications
4. Attention foci

The logical connective can be OR, XOR, or AND, together with NOT, and contains an action which will occur if the goal beneath it is achieved. Currently the action is a modification to the PRIORITY value or KILL. Functional group modifications are requests for interchange or introduction of a functional group at a particular location on the target, or the converse, to leave a group unchanged. Structural modifications currently state bonds to be made or broken or the converse. Attention focussing instructions require a transform to use specified atoms, bonds, or groups or the converse, don't use the specified item. A goal is a logical connective joining a list of one or more of the other entities which may include another goal. The longifolene example would be

goal 1: (OR, BREAK 1-7, BREAK 3-4, GOAL 2)
goal 2: (AND, BREAK 3-4, BREAK 1-2).

The strategy module can be interrupted prior to selection of transforms, but after the relevant strategies have set up their goals. The goal list can be printed out, new goals added, or existing goals modified or deleted by the user. This interactive capability to add goals facilitates testing strategies and allows the user to supply his/her own strategies if they are not included in the SECS strategy module. The goal list can be used to select only those transforms satisfying the goals or it can be used to simply raise the priorities of those satisfying the goals. Note that the goals may not refer by name to any transform, instead goals are strictly defined in terms of the target structure.

7.5 TRANSFORM SELECTION by relevance to the goal list requires that the goal list interpretor have some knowledge of what kinds of things a transform might do. This is described generally by the transform CHARACTER (see section 3) which contains any number of the phrases given below. Thus a transform whose sole character was ALTERS GROUP would not be considered relevant to a goal of breaking a carbon-carbon bond. The transform character descriptions are collected together in a directory for rapid determination of possible relevancy to the goals. If the character is suitable then the SUBSTRUCTURE of the transform is examined since it also helps define the character of the transform. At this point if the transform still appears to be

BREAKS CHAIN, RING, 3RING, 4RING, or 6RING
MAKES BOND, 3RING, 4RING
INCREASES or DECREASES CHAIN
EXPANDS or CONTRACTS RING
INTRODUCES or REMOVES GROUP, STEREO, or AROMATICITY
ALTERS GROUP
INVERTS/RETAINS STEREO
MIGRATES BOND

relevant, the transform is checked for applicability, i.e., the SUBSTRUCTURE is compared to the target. If the transform "fits," then it is interpreted in detail as described in section 3.1. The transform manipulation instructions are compared to the goal list to determine if the transform actually was relevant, and appropriate actions are taken. The transform may be aborted, in which case no precursor is ever generated. When the target molecule contains elements of symmetry the goal interpreter also checks to see if the action of a transform is equivalent by symmetry to a specified goal although the actual atom numbers may be different.

7.6 SUBGOAL GENERATION. If the SUBSTRUCTURE does not "fit" the target, yet the transform appears relevant, a subgoal is generated to correct the mismatch so the transform will "fit." The technique of generating subgoals as a result of mismatch was even used in the very first synthesis program.[6] The difference between the required SUBSTRUCTURE and the target is used to create a subgoal, normally of the functional group modification type. This is an example of decomposing a problem into subproblems, and approaching each subproblem by Means-Ends analysis.[27] The result of this goal-directed approach is that the troublesome FGI and INTRO transforms are only applied when they are needed in order to enable application of another transform relevant to the goal list. This is an extremely efficient method for reducing the number of poor pathways generated without losing any good pathways. Subgoals are identical in form and function as goals, but are kept on a separate list and become active only after the original goal list relinquishes control. Subgoals are not created explicitly by transforms, thus the transform writer need never consider what possible subgoals might be, or where they may be invoked. This is another aspect of our efforts to maintain complete separation between strategies and transforms.

8. CONCLUSION

We have discussed two important methods used in SECS for improving the chemical accuracy and efficiency of syntheses generated in the synthesis tree. The first method involves improving the accuracy of chemical inference through analysis of various models and through accurate scope and limitations in the transforms. As illustrations we described the evaluation of steric effects, electronic directing effects, and interfering functional groups including prescription of protecting groups.

The second method involves selection of transforms by relevance to goals created by synthetic strategies rather than selection by applicability. This was made possible by the development of a uniform representation of goals as dynamic logical list structures describing desired changes in the target molecule, and by including in each transform a description of the transform's character. Of course involving the user-chemist in decisions concerning which precursor to process next is another important method, fundamental to our interactive approach. Additionally we stressed the importance of separating strategies from transforms, not only to simplify the addition of new transforms, but also to minimize bias and maximize creativity in the synthetic sequences generated.

ACKNOWLEDGEMENT. This work was supported in part by NIH grant RR00578, and by an allocation of the SUMEX-AIM resource at Stanford (RR00785, J. Lederberg, principal investigator). The authors also wish to thank IBM and Merck, Sharp and Dohme for fellowships to G.S., the CNRS for a fellowship to F.C., the Deutsche Forschungs-gemeinschaft for a fellowship to H.B., and Sandoz, Basle for support of W.S. We also gratefully acknowledge the contributions of Dr. S. Krishnan to the goal list editor and interpreter and of Prof. Clark Still to the first version of the goal list interpreter.

References

1. D.B. Lenat, "AM: An Artifical Intelligence Approach to Discovery in Mathematics as Heuristic Search," Ph.D. Thesis, Computer Science Dept., Stanford University, 1976.

2. W.T. Wipke, "Computer Planning of Research in Organic Chemistry," Proceedings of Third Intl. Conf. on Computers in Chemical Education, Research and Technology, Plenum Press, July 1976.

3. W.T. Wipke, "Computer-Assisted Three-Dimensional Synthetic Analysis," in Computer Representation and Manipulation of Chemical Information, ed. W.T. Wipke, J. Wiley, 1974, pp. 147-174.

4. T.M. Gund, P.v.R. Schleyer, P.H. Gund, and W.T. Wipke, J. Am. Chem. Soc., 97, 743 (1975).

5. This is in contrast to the FIEM of Ugi described elsewhere in this volume.

6. E.J. Corey and W.T. Wipke, Science, 166, 178 (1969).

7. E.J. Corey, W.T. Wipke, R.D. Cramer III, and W.J. Howe, J. Am. Chem. Soc., 94, 421, 431 (1972).

8. R.J. Feldmann, "Interactive Graphical Chemical Structure Searching," in Computer Representation and Manipulation of Chemical Information, ed. W.T. Wipke, J. Wiley, 1974, pp. 55-82.

9. W.T. Wipke and T.M. Dyott, J. Chem. Info. and Computer Sci., 15, 140 (1975).

10. W.T. Wipke and T.M. Dyott, J. Am. Chem. Soc., 96, 4825, 4834 (1974).

11. W.T. Wipke and H. Braun, submitted for publication.

12. W.T. Wipke, P. Gund, T.M. Dyott, and J.G. Verbalis, in preparation.

13. W.T. Wipke and P. Gund, J. Am. Chem. Soc., 96, 299 (1974); 98, 8107 (1976).

14. W.T. Wipke, T.M. Dyott, C. Still, and P. Friedland, in preparation.

15. W.T. Wipke and P.H. Gund, J. Am. Chem. Soc., 96, 299 (1974); 98, 8107 (1976).

16. W.T. Wipke and P. Friedland, unpublished.

17. E.J. Corey, W.J. Howe, and D.A. Pensak, J. Am. Chem. Soc., 96, 7724 (1974).

18. H.H. Jaffe, Chem. Rev., 53, 191 (1953).

19. J.B. Hendrickson, J. Am. Chem. Soc., 93, 6854 (1974).

20. G.W. Wheland, J. Am. Chem. Soc., 64, 900 (1942).

21. Physical Methods in Heterocyclic Chemistry I, Academic Press, N.Y., 1963, p. 140.

22. A. Streitweiser, Jr., Molecular Orbital Theory for Organic Chemists, J. Wiley, N.Y., 1961, p. 117.

23. L. Velluz, G. Nominé, J. Mathieu, Angew. Chem., 72, 725 (1960).

24. W.T. Wipke, T.F. Brownscombe, G. Birkhead, and P. Gund, in preparation.

25. E.J. Corey, M. Ohno, P.A. Vatakencherry and R.B. Mitra, J. Amer. Chem. Soc., 86, 478 (1964).

26. U. Biethan, U.v.Gizycki, and H. Musso, Tetrahedron Letters, 1477 (1965).

27. G. Ernst and A. Newell, GPS: A Case Study in Generality and Problem Solving, Academic Press, N.Y. 1969.

28. P. Karrert and C.H. Eugster, Helv. Chim. Acta., 33, 1172 (1950).

29. H.H. Inhoffen, F. Bohlmann, K. Bartram, G. Rummert, H. Pomner, Ann., 570, 54 (1950).

30. N.A. Milas, P. Davis, I. Belic, and D. Fles, J. Am. Chem. Soc., 72, 4844 (1950); See also Carotenoids, Otto Islor, ed., Birkhaüser Verlag, Basel, 1971.

6

Rapid Generation of Reactants in Organic Synthesis Programs

MALCOLM BERSOHN
Dept. of Chemistry, University of Toronto, Toronto, Canada M5S 1A1

The question of efficiency of reactant generation has not received primary attention and this is as it should be since in a problem requiring more than about four steps it is more important to develop better heuristics to restrain the generation of reactants than it is to find ways to generate the latter more rapidly. Furthermore, the program of the future may well spend as much of its time searching an external data base as it does generating reactants. The external data base would be, in essence, the synthetic part of Chemical Abstracts. It would contain the solutions of standard problems so whenever the molecule at hand is recognized to be similar to a standard problem then the stored solution, a sequence of reactions, is retrieved. In this future situation the representation of molecular structure used internally by the synthesis program may have to be the same as that of the external data base. Hence the molecular structure representation might have to be chosen primarily from this point of view rather than to optimize the speed of reactant generation. All this being said, reactants still have to be generated and the rapid generation of reactants is an economic benefit. Speeding up the generation of reactants means speeding up the component routines. Of these the most time consuming are 1, canonicalization of the molecular structure representation, 2, finding the rings, 3, finding the functional groups and 4, retrieving the reactions and performing the tests to decide whether the particular product molecule at hand is suitable as a product of the reaction. Comparatively speaking, the actual generation of the structure of the reactant molecule(s) is a brief operation. In my programs the most time consuming single routine is the canonicalization of the molecular structure representation and therefore approaches to the acceleration of this routine will be discussed first.

I. Canonicalization of the Molecular Structure Representation

Basically, canonicalization consists of numbering the

non-hydrogen atoms of the molecule according to a set of rules. This means that a molecule can be represented in only one way. Having numbered the atoms according to the rules we are rewarded with several advantages, namely:

1. We can speedily recognize whether the molecule at hand is the same as a previously generated reactant molecule or the same as a molecule known in the program as being available. Without canonicalization we would be forced to do some kind of atom by atom matching (1) in order to determine if two structures are the same.

2. In the course of deciding the precedence of the atoms we necessarily have to discover any equivalence between atoms that exist in the molecule. Atoms are said to be equivalent if they are carried into each other's position by global or local symmetry operations of the molecule. Global symmetry operations are rotations or reflections or combinations of these with respect to infinitely long axes or infinite planes passing through the center of the molecule. Local symmetry operations are rotations about bounded bounded axes and reflections in bounded planes. In the figure below we see a molecule

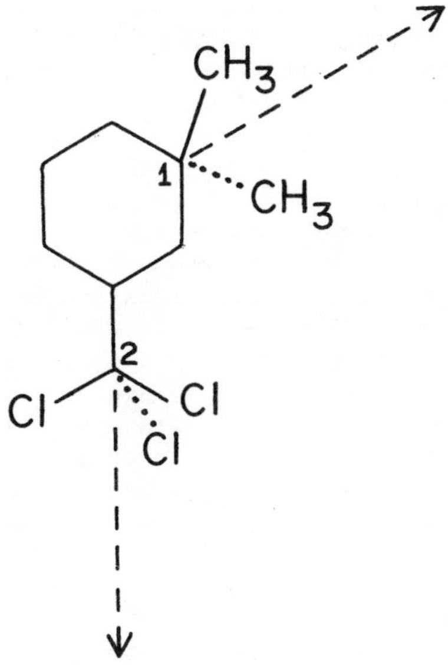

1,1-Dimethyl-3-trichloromethylcyclohexane, which has local symmetry. A local C_{3v} axis terminates in atom 1 and a local C_2 axis terminates in atom 2. (The point group of the molecule is the direct product group $C_{3vL} \otimes C_{2L}$, where the suffix L means local.) The equivalence of the chlorines and the two methyl groups can be discovered by canonicalization without having to build a model and actually perform a local reflection or rotation to see if the result is the same molecule. Similarly, with molecules that have global symmetry, such as methylcyclohexane, the presence of two pairs of equivalent methylene carbon atoms can be found without performing a global reflection. Knowing which atoms are equivalent to each other is necessary for concluding that chirality is absent in the common ligand of equivalent atoms. Under various conditions, many reactions produce two products in significant yield and are therefore not to be recommended when a certain pair of reacting atoms are not equivalent. When this pair is equivalent the reaction produces one product and the reaction is to be recommended. It is therefore absolutely necessary for a synthesis program to know which atoms in the molecule at hand are equivalent. The reactions include electrocyclic reactions, reactions involving the carbon atom alpha to a ketone when both alpha atoms have the same number of attached hydrogen atoms, Wittig type reactions etc.

3. Having numbered the atoms canonically, these numbers provide a global ordering of the atoms which can be used locally at a chiral center to determine whether the center should be called R or S. Thus, if the ligand atoms of a chiral atom are canonically numbered 3,7,9 and 14 and atoms 3,7,9 are arranged in a counterclockwise fashion when viewed from the side opposite atom number 14 then we can mark the atom with an S. This internal R,S notation may differ for some centers from those provided by the Cahn-Ingold-Prelog procedure (2) but there is no difficulty in translating between the systems since the absolute configuration is embodied in the molecular representation. (Normally this will not be necessary as most reactants never see the light of day: in my noninteractive synthesis programs, tens of thousands of molecular structures are often generated before an acceptable synthetic pathway is produced.) Thus we are spared the trouble of determining the sequence of the four ligands in the Cahn-Ingold-Prelog sense.

4. If the atoms are numbered, other things being equal, in order of their atomic number and thereafter in order of their degree of unsaturation, then choosing the descending orders, oxygen atoms precede nitrogen atoms which precede carbon atoms, unsaturated oxygen atoms precede saturated oxygen atoms etc. Now if we further order the list of the ligands of each atom in ascending order of the numbers of the atoms then it is often possible for a subprogram to know "in advance" which ligands are which. For example, if a program is examining the description

of the nitrogen of an amine oxide, then the first number encountered in the list of ligands of the nitrogen is that of the oxygen atom and the next number is that of the carbon atom ligand. Again, if the program is examining an ester carbonyl carbon, the first ligand encountered in the list will automatically be the unsaturated carbonyl oxygen, the next ligand will be the saturated ether oxygen and the last ligand will be the alpha carbon atom. The program can pick up these numbers and use them in other contexts without examination of the data about the atoms to which the numbers refer.

Accepting that the application of a set of rules for numbering the atoms of each molecule considered by the program is necessary, one might ask why not use the book full of IUPAC rules? (3) The problem here is that since the IUPAC numbering rules depend upon the ring system being considered, the programming of this book full of special cases is an enormous task, not worth while. It is much easier to have briefly stated rules.

Computerized Procedures for Numbering the Atoms of a Molecule

W.T. Wipke (4) first achieved the canonicalization of a molecular structure representation in a computer program that includes all aspects of stereochemistry. In some other schemes, stereochemistry is not invoked to aid in the numbering of the atoms. Such schemes are incomplete. If we relegate stereochemistry to a footnote and the numbering of the atoms and the connection tables of two isomers can be the same, then we lose most of the above-stated advantages of canonicalizing.

H. Gelernter's program (5) is unique in that the sorting is done via the Wiswesser line notation. The Wiswesser scheme (6) sorts the groups and the numbering of the atoms follows from their positions in the groups and the order in which the groups are given in the Wiswesser symbols that convey the structure of the given molecule. All other schemes reported in the literature (7,8,9,10) for canonicalizing a molecular structure representation require a direct ordering of the non-hydrogenic atoms of the molecule. In what follows we will omit the modifier "non-hydrogenic" and ask the reader to note that the hydrogen atoms are not numbered but are considered to be properties of their ligands. We can divide the methods already in use into tree algorithms and sum algorithms, depending on how the external environment of each atom is represented.

The atoms of a molecule can be partitioned into equivalence classes on the basis of a single property or a set of properties. The value of each class for the property(s) will be called the initial canonical value. The set of properties could include the atomic number, predominant in the Cahn-Ingold-Prelog system if this is used to number the atoms of the molecule, or the number of ligand atoms, which is the property of the greatest use in the Morgan algorithm. We can also include the degree of unsaturation, the number of attached hydrogen atoms, the charge and information about the size of the ring(s) of which the atom is a member. It is evident that the more of these pieces of information that comprise the initial canonical value, the more equivalence classes we will have. Thus if we use only the atomic number we have a large equivalence class with the value 6. If we add to this a number representing the degree of unsaturation, the number of equivalence classes is increased and the size of each class is reduced. Still there may be many atoms with the initial canonical value 6.0, i.e. saturated carbon atoms. There are many others, perhaps with the value of 6.1, aromatic carbon atoms or 6.2, doubly bonded carbon atoms, etc. If we further include the number of attached hydrogen atoms we can have the equivalence classes of 6.0.2 and 6.0.1 referring to saturated methylene and methinyl carbon atoms respectively. Adding the ring size we then distinguish the class with the initial canonical value 6.0.2.5 from the class with the initial canonical value 6.0.2.6.

Different schemes of canonicalization can be distinguished by the properties selected to establish the initial canonical value and whether the ligands of each atom are characterized by a tree of such initial canonical values or the sum of such values obtained by a successive summation process to be detailed later.

Tree Algorithms and Sum Algorithms

We illustrate the tree approach with an example, using the molecule of Figure 1. In this figure the atoms are numbered arbitrarily,

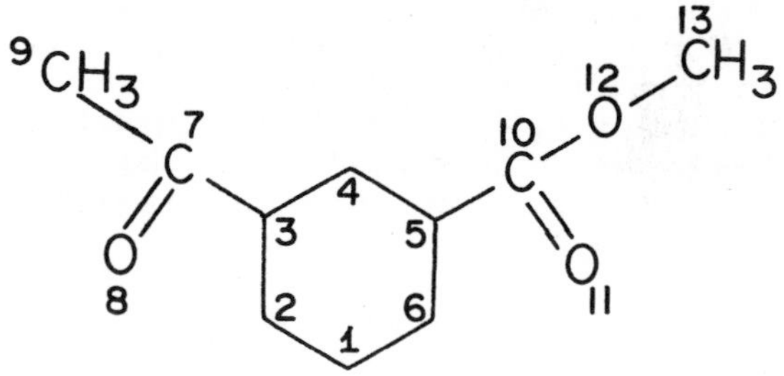

Figure 1. A molecular structure with arbitrarily numbered atoms

for reference in the discussion, not canonically. We will take as the set of canonical properties the single property of atomic number, and treat double bonds as meaning the double occurrence of the atom concerned, both as in the Cahn-Ingold-Prelog system. We consider atoms 2 and 6. The initial canonical values for these are both 6. Hence in the effort to distinguish them, we walk out in all directions and compare the canonical values along the paths from both the atoms. The paths of length one, i.e. taking account only of the ligands, give strings of 6.6 for both atoms. The paths of length two give strings of 66.66.66 for both atoms. The paths of length three give strings of 668.668.666.666.666 for atom 2 and 668.668.668.666.666 for atom 6. Hence we can conclude that atom 6 outranks atom 2. The tree algorithm terminates under the following various conditions: 1. No two atoms have identical trees. 2. The pairs of trees describing the environment of all pairs of atoms whose equivalence is in doubt either converge to a common atom or else involve every other atom of the molecule. In a real situation, at this point a tree of stereochemical values has to be built if there are equivalent atoms in the molecule.

Now let us examine the behaviour of the corresponding sum algorithm. Here we will use the same initial canonical value. The second canonical values are the sum of the ligands' initial canonical values. In general the ith canonical value for an atom is the sum of the i - lth canonical values of its ligands. In this way we obtain a second canonical value of 12 for both atoms 2 and 6 of Figure 1. The third canonical value is 30 for both atoms. The fourth canonical value is 76 for atom 2 and 78 for atom 6. Thus it is on the third iteration of the summing process that the canonical value for atom 6 finally receives the information that atom 12 is an oxygen atom. This information is not conveyed directly but it is mixed into the sum along with the properties of other atoms. The sum method terminates when

	Canonical Value Number			
Atom row number	1	2	3	4
1	6	12	24	60
2	6	12	30	76
3	6	18	52	114
4	6	12	36	108
5	6	18	54	102
6	6	12	30	78
7	6	28	48	192
8	8	12	56	96
9	6	6	28	48
10	6	30	54	212
11	8	12	60	108
12	8	12	38	66
13	6	8	12	38

the number of equivalent atoms cannot be reduced between two successive iterations (method of refs. 9,10) or when the atom of highest connectivity has been found (Morgan's algorithm). In a real case, if there are equivalent atoms in the molecule then sums involving stereochemistry have to be computed. The advantage of the sum method is ease of computation. It is easier to compare simple numbers than long strings of numbers. It is also faster to generate the sums.

The canonicalization procedure of Professor Ugi's group is a tree algorithm, in which the initial canonical value is composed of the atomic number and the coordination number of the atom. (If one of these two numbers makes the atom unique in the molecule then only the one number is used as the initial canonical value.)

Consider the molecule partly depicted below, in Figure 2.

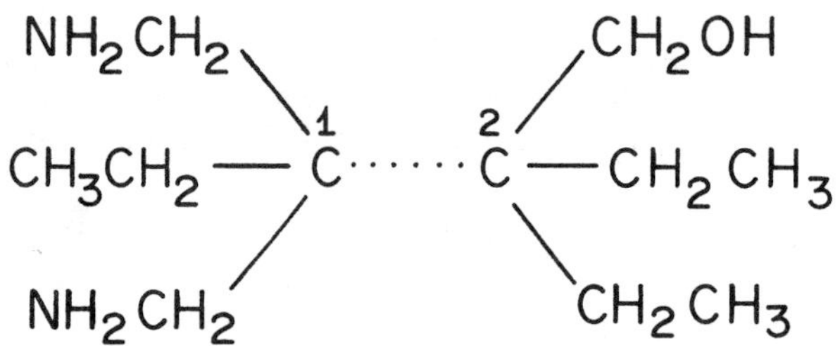

Figure 2

By the dotted lines we will mean any string of atoms without side chains which are equivalent with respect to atoms 1 and 2. It is clear that atoms 1 and 2 are nonequivalent. But they will send out sums of various categories which are the same. The sums of the unsaturations are all zero, all the atoms concerned are acyclic, the sum of the atomic numbers are the same as well as the sum of the number of the attached hydrogen atoms. Morgan's algorithm on encountering a "tie" like this would examine the ligands to decide which atom should be chosen, according to its rules, as the atom to receive the lowest number. The beginning atom is the only one which is numbered because of the precedence of its final canonical value. Other atoms are numbered according to their relation to it. The initial canonical value is the number of ligands. The summing part of the algorithm terminates when we can no longer increase the number of equivalence classes. There is no attempt to use the canonical values themselves as the basis for ordering all the atoms. In my algorithm the final canonical values are used as the criterion for determining the precedence of atoms with identical initial canonical values.

Other Algorithms Not Classifiable as Tree or Sum Algorithms

1. The Repeated Renumbering Algorithm

It is possible to achieve the purpose of a tree algorithm without directly comparing strings. We number the atoms according to their initial canonical values, using an arbitrary numbering for atoms with the same initial canonical value but making sure that if $i<j$ then the initial canonical value of atom i is greater than or equal to the initial canonical value of atom j. Next we renumber the atoms, ranking them first on the basis of their initial canonical values and then on the basis of the previous numbering of their ligands. We keep repeating the renumbering process until the connection table or other molecular representation is unchanged by the renumbering. This method appears elegant but in our hands it has always been slower than a sum method. The renumbering is extensive in the first few iterations and the consumption of time in translating between successive numberings proved excessive. This algorithm still seems worthy of further investigation. It is possible that improved programming of the subalgorithms can markedly improve this scheme.

The example below shows the first iteration of this algorithm when used to order the atoms of the molecule shown

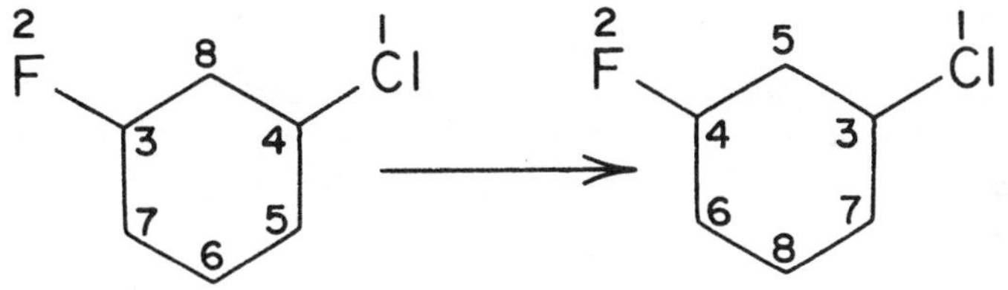

2. Canonicalization Based on Atomic Coordinates

This idea involves first numbering the atoms according to their initial canonical value. Atoms having the same canonical value would have their relative precedence decided by their distances from the center of mass of the molecule. The distance would be in nanometers, not bonds, so there would have to be a uniform way of building the model and deciding on the average or most stable conformation. This method has the merit of doing away with the need for iterations, whether they be iterations of the summing process or iterations of the tree growing process. On the other hand the discovery of equivalences of atoms under local symmetry operations presents a problem.

A Sum Algorithm Described in Detail

My programs use a sum algorithm to be described below. The time for canonicalizing the steroid 6α-methyl-17α -acetoxyprogesterone is 5063 times the time for executing a store instruction on the same computer, the IBM 370/165. (The actual time is 1.61 + 0.02 microseconds. I deliberately place this number in parentheses, since all such comparisons should be relative to the computer.) A search of the literature did not reveal other such reported times, so there is no reason to conclude that the sum algorithm discussed below is less or more rapid than any other canonicalization algorithms used elsewhere. Hopefully this paper presents some starting points for considering the question of efficiency and how to improve it. The reactant generation time for difficult problems, e.g. steroids, is 4-5 milliseconds, hence the programs spend as much as 40% of their time canonicalizing the molecular structure representation. Evidently, here is the most important place to think about efficiency.

It is necessary first to present explicitly the molecular structure representation used. The structure is described by two tables. The first is a connection table; the second is a stereochemical relation table. Connection tables and their alternatives are well discussed in the book of M.F. Lynch et al. (11). Both tables have to be canonicalized; in both tables the ith row describes atom number i. The connection table contains 12 columns or fields that describe properties of atom i and these are followed by columns 13-16 in which are listed the row numbers of the atoms which are ligands of atom i. (For the bit oriented reader, I explain that the first 12 columns occupy 32 bits and the last four columns also occupy 32 bits. This imposes a limit of 254 non-hydrogen atoms in the molecule: the value 255 is reserved to mean a blank.) The first twelve columns of the connection table are as follows: 1. the atomic number, 2. the degree of unsaturation, 3. four minus the number of hydrogen atoms, 4. the elementary functional group serial number, 5. member of a ring of size other than 3,5, or 6, 6. member of a ring of size 3, 7. member of a ring of size 5, 8. member of a ring of size 6, 9. positive charge, 10. negative charge, 11. R chirality, 12. S chirality.

The connection table of 2-Butanone-3-ol (R) is given on the next page.

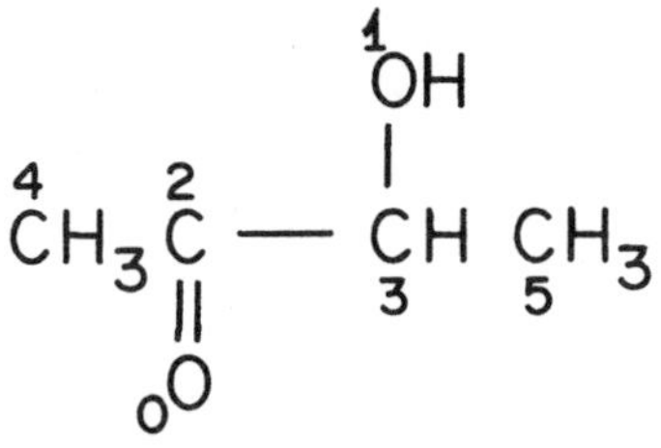

1 OH

4 2 |

CH_3C — CH CH_3

|| 3 5

0 O

atomic number	unsaturation	four minus # H atoms	elementary group	ring	+-RS	Ligands
8	4	4	0	0000	0000	2 2
8	0	3	0	0000	0000	3
6	4	4	19	0000	0000	0 0 3 4
6	0	3	7	0000	0000	1 2 5
6	0	1	0	0000	0000	2
6	0	1	0	0000	0000	3

The "degree of unsaturation" is 0 for saturated atoms, 1 for aromatic atoms, 2 for carbon atoms doubly bonded to carbon atoms, 4 for atoms involved in a double bond to a heteroatom, 6 for ketene carbonyl carbons, 8 for triply bonded atoms.

Atoms which are in some way central to an elementary and heteroatom-containing functional group, such as the oxygen of an ether or the carbonyl carbon of a ketone etc. are labelled with a number that corresponds in our program to that elementary functional group. These numbers, listed in Table I are fundamental to the discussion in part II of this paper. Here we observe that the value for column 4 is zero for all atoms which are not central to a heteroatom-containing elementary group of Table I. Going back to Figure 2, we note that the canonicalization system used here cannot regard the atoms 1 and 2 as being equivalent since the initial canonical values include the values of column 4. The sum of these latter values for atom 1 is different from that for atom 2.

Columns 5 through 12 contain either the value 0, meaning the atom lacks the relevant property or 1, meaning the atom posseses the relevant property.

The stereochemical relation table contains in the ith row the steric relations of atom i to other atoms. The entries in the table consists of pairs of symbols, the first indicating the steric relation and the second indicating the atom concerned. 1.9 means that atom i is axial in the ring of which this ligand atom 9 is a member. 2.9 is the corresponding equatorial indication. 5.6 would mean that atom i is a substituent of a double bond and is cis to atom 6. 6.8 would mean that atoms i and 8 are trans across some double bond. The symbols 7 and 8 refer to cis and trans relationships with respect to a ring. The full list of symbols is given in reference 9.

The initial canonical value is based on the values of all of the first 10 columns of the connection table. Successive canonical values are sums of the values of these ten properties, including data from successively remote atoms. (To avoid overflow of, for example, the third column onto the second column, the ten numbers may be expanded into numbers with more digits and the ensemble can be kept in a 64 bit number manipulated by floating point registers. This time consuming expansion can be avoided by following the procedure of the appendix, making sure to include step 6.) Evidently the larger the set of properties from which the initial canonical value is derived, the more

Table I. Serial Numbers of Elementary Functional Groups

```
 1 add one for CH
 3 add 3 for phenyl to make benzylic derivative
 5 CH2OH primary alcohol
 7 CHOH secondary alcohol
 9 COH tertiary alcohol
11 sulfoxide
13 sulfone
15 nitroseo
17 nitro
19 ketone
21 ester methyl or other trivial ester
23 aldehyde
25 nitrile
27 dithiane
29 carboxylic acid
31 ketal
33 acetal
35 hemiacetal
37 ether
39 epoxide
41 CI
43 hydrazine
45 primary amine
47 secondary amine
49 tertiary amine
51 imine
53 urea
55 primary amide
57 secondary amide
59 tertiary amide
61 CCl
63 thiol
65 C-S-C sulfide
67 C-S-S-C disulfide
69 CF
71 ketene; the middle carbon is a central atom
73 bromide
35 acyl halide
77 carbodiimide
79 isocyanate
81 amine oxide
83 oxime
85 substituted hydroxylamine
87 azo C-N=N-C
89 sulfonic acid
91 sulfonamide
93 sulfinic acid
95 mesylate
```

Table I. Continued

```
 97 C=C; add 97 to C-X serial number to obtain serial # of
    allylic C-X
 99 C=CH
101 CH=CH
103 C=CH2
105 CH=CH2
107 cyclohexene
109 C-C triple bond
111 phosphine oxide
113 phosphinic acid
115 anhydride
117 imide
119 primary ester CO2CH2R
121 secondary ester CO2CHRR
123 tertiary ester CO2CRRR
125 primary phosphine
127 secondary phosphine
129 tertiary phosphine
131 add 131 to C-X serial # to obtain aromatic C-X serial #
133 secondary borane
135 tertiary borane
137 peroxide
139 hemiketal
141 enol lactone
143 vinyl halide
145 primary vinyl amine
147 secondary vinyl amine
149 tertiary vinyl amine
151 cyclopropane ring
153 cyclobutane ring
155 allene middle carbon is central atom
194 diene, a complex functional group, (97 + 97)
```

distinctions can be made between atoms. Such distinctions include the distinction between ring atoms and acyclic atoms, between an ester carbonyl carbon and a ketone carbonyl carbon etc. The more such distinctions can be made the fewer iterations of the summing process are required. A central idea here is that a synthesis program, unlike a program processing manual input, knows all about the molecule from which the molecule at hand is derived hence matters like what size ring if any the atom is situated in or what functional group the atom is central to, are data available to a synthesis program before canonicalization of the molecular structure representation.

When we procede from the kth to the k + 1th canonical value and there is no decrease in the number of equivalent atoms, then the calculation of the canonical values terminates. All atoms are then numbered in the order of the value of the first 10 columns of the connection table. Whenever these are equal the final canonical values are consulted to determine which atom precedes which. If the final canonical values are the same we have good grounds to suspect that the atoms are equivalent but they must meet two more requirements to be judged equivalent:

1. If there are at least two nonterminal double bonds in the molecule the algorithm continues and calculates a "stereochemical canonical value", in which the initial canonical value is a number meaning cis or trans with respect to a double bond. If the atoms concerned are still equivalent with respect to the steroeochemical canonical value, then they are truly equivalent unless they are both chiral. This step is required to reveal the inequivalence of atoms 1 and 2 of the Figure below.

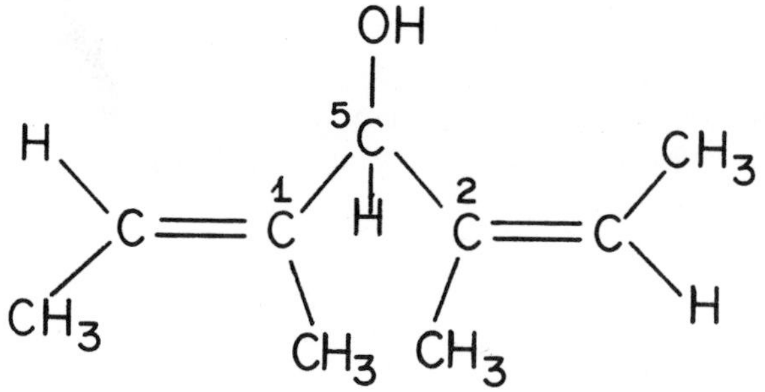

2. If the atoms are chiral the program must determine the chirality and if they are both R or both S they are actually not equivalent. (cf. reference 9).

If, let us say, atoms 9 and 10 are equivalent, then which should be numbered 9 and which should be numbered 10? How, for example, should we number the atoms of cyclohexane? The answer is that if i<j then the row numbers of the ligands of i which are not j should be less than or equal to the row numbers of the ligands of j which are not i. Application of this rule gives 1,2,4 and 3 for the numbering of the atoms of cyclobutane, reading around the ring. Similarly, reading around the ring, this rule gives 1,2,4,6,5 and 3 for the atoms of benzene. We require that the ligand row numbers of i and the ligand row numbers of j being compared must not include i or j. If we drop this stipulation then we would have an infinite loop in the case of ethane or symmetrically substituted ethanes or rings of equivalent atoms.

We return once more to Figure 2. Thanks to the presence of column 4 the procedure here cannot regard the two atoms as being equivalent. But what if a more complicated situation arose such that two atoms which are not equivalent had the same ultimate canonical value? I consider this most unlikely and have not been able to find such an example but in any case if this is encountered it can easily be managed by including in the algorithm the requirement that equivalent atoms must have pairwise equivalent ligands. On detecting the nonequivalence of the ligands the precedence would follow that of the ligands. (Inclusion of this requirement removes the need to make sure that the various columns from the initial canonical value do not overflow onto columns to the left of them. In the very rare case in which inequivalent atoms appear to be equivalent because of such overflow, the atoms will be detected as inequivalent by the above stated requirement that equivalent atoms must have equivalent ligands.)

It is necessary to order the ligand atoms in column 13-16. We happen to use ascending order. In the process of sorting the row numbers of the four ligand atoms it is easy to detect inversions of chirality that have taken place in a reaction. If an even number of exchanges is required to order the numbers of the ligand atoms, then the chirality of the atom is the same in the reactant as in the product. If an odd number of exchanges is required to order these row numbers then the chirality of this atom in the reactant is opposite to that of the atom in the product. If two atoms of the reactant are equivalent any common ligand cannot be chiral even if it is chiral in the product so a zero is placed in columns 11 and 12. Similarly if a saturated carbon or quaternary nitrogen atom has no two equivalent ligands and the atom was not chiral in the product then the chirality is a problem to be determined, using the methods of reference 9. The previous manipulations of chirality are included in the time required for canonicalization of the connection table but the problem of deciding chirality when the atom is not chiral in the product is handled by a different routine.

II. The Numbering of Synthetically Interesting Substructures

and the Use of These Numbers for Rapid Discovery of Functional Groups

Most synthetic reactions produce a well defined substructure in the product. It is therefore natural to classify the reactions by the substructure produced in the product. This is done in the standard compendia such as those of Buehler and Pearson, (12) and Harrison and Harrison. (13) It is also natural for a programmer to number these substructures and use these numbers as indices for retrieval of reactions that produce this substructure. The problem of recognizing the substructures becomes one of deriving these numbers in some way from the molecular structure representation. In the representation which I use, the numbers characterizing elementary functional groups are built into the representation so the derivation of the elementary functional group numbers is an immediate extraction. (14, 15) As mentioned above, column 4 of the connection table has the value zero except for atoms which are central to some elementary functional group that contains a heteroatom. Elementary functional groups which do not contain a heteroatom, i.e. carbon-carbon multiple bonds, are easily collected from the contents of column 2, the degree of unsaturation, which places such atoms ahead of all other carbon atoms. A loose definition of an elementary functional group is that it is a carbon-carbon multiple bond or a heteroatom and its carbon ligand(s) together with their ligands. A more practical definition of an elementary functional group is that it is any functional group that appears in Table I! This latter definition reveals the empirical and arbitrary nature of this classification. Complex functional groups are sets of atoms in which we can find more than one elementary functional group.

As a simple example, we can look at the connection table of 2-butanone-3-ol given above. As we scan column 4 we see that atom 2 has the label 19, meaning ketone and atom 3 has the label 7, meaning secondary alcohol. There are no other elementary functional groups in the molecule, since, as shown in column 2, there are no vinyl or acetylenic carbons.

Having our list of elementary functional groups we then look for alpha, beta, gamma, delta and epsilon relationships between them. This search reveals that the ketone and secondary alcohol are adjacent, giving us the complex functional group with serial number 26, the sum of 19 and 7, and alpha ketol. Complex functional groups are thus found starting from the elementary functional groups. The numbering of complex functional groups is not one to one since, for example, there are other ways of forming the number 26 from the sums of two odd integers. Consequently future use of the functional group has to be preceded by a test on the first central atom to ascertain if it is a secondary alcohol. The format of the functional group representation in a list of functional groups prepared by the program is: serial number of the functional group, row number of the first central atoms, row numbers of the ligands of the first central atom, row number of the second central atom, row numbers of the ligands of the second central atom, row numbers of other relevant atoms. An elementary functional group with a heteroatom has only one central atom. A carbon-carbon multiple bond is written so that the carbon with fewer attached hydrogen atoms is in the position of the first central atom and the number of the other atom is in the position of the second central atom. Below we show the functional group list of 2-butanone-3-ol.

```
26  2  0  0  4     3  1  5
 7  3  1  2  5
19  2  0  0  3  4
```

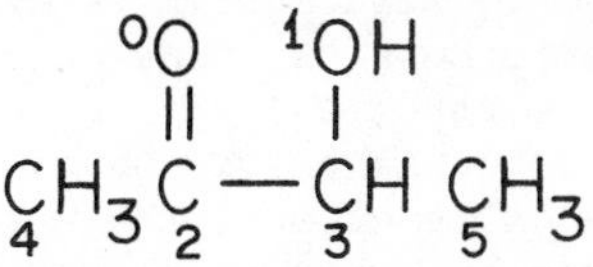

The ambiguity of the numbering of the complex functional groups in which a test must be performed on column 4 of the first mentioned atom before the identity of the complex functional group is certain, is the result of a temporary compromise with the hardware. A truly high speed virtual storage will enable the complex functional groups to be numbered unambiguously, so that a tertiary alcohol next to an ester will be numbered differently from the complex functional group consisting of a secondary alcohol next to a ketone. These unambiguous numbers will be larger than the present ones.

The number of atoms intervening between the central atoms of the two functional groups is given as a prefix to the number of the complex functional group. For example, an alpha keto ester is numbered 0.40, a beta ketoester is numbered 1.40, a gamma ketoester 2.40 etc.

It is unnecessary to find all the functional groups of every reactant molecule. Many reactions simply alter, insert or remove one functional group. A suitable flag posted on the description of such reactions should alert the synthesis program that the reaction is of this type and the full functional group finding routine need not be called. Instead the functional groups are copied from the product to the reactant, with suitable renumbering of the atoms involved. Next the alpha, beta gamma etc. relations of the new or different functional group with the other functional groups are looked for. If the reaction consists of inserting a new functional group then the functional group in question is simply removed from the list of functional groups copied from the product. Failure to use such short cuts will add noticeably to reactant generation time.

Some substructures of synthetic interest are neither rings nor functional groups but are just hydrocarbon fragments. For example, if the reaction is desulfurization of a 2-substituted thiophene derivative then the product contains a string of four saturated carbons, which are originally part of the thiophene ring, along with the removed sulfur atom. The predominant theme of the author's approach to substructure discovery is that the data should be structured so that the group of interest should simply be collected and not searched for specifically or inquired about by questions in some logical sequence that narrow down the possibilities. But in this case I cannot fina a way to avoid a specific inquiry, and hopefully the ingenuity of others will provide some solution to this kind of problem.

An even more important problem is the problem of discovering what I will call hyperstructures. Typically they will involve ring systems and/or three or more functional groups. What has to be retrieved, once these are discovered, is not an ordinary synthetic reaction but a highly specialized reaction, perhaps using an enzyme or microorganisms or else a specialized sequence of reactions, the performance of which will produce the discovered hyperstructure in good yield. The recognition of such hyperstructures will probably replace, to a noticeable extent, the search process of generating reactants.

The discovery of such hyperstructures procedes in our subprogram as follows. If one has discovered a constituent part of this hyperstructure, a subroutine is called which inquires about various hyperstructures of which this could be a part. Thus the discovery of fused aromatic rings leads to the inquiry sequence for finding phenanthrene. An ordinary aromatic molecule would not be quizzed for the phenanthrene possibility. Similarly, the discovery of fused nonaromatic rings leads to questions searching toward the existence of a steroid skeleton. An example is the 11-alpha-hydroxylated trans-anti-trans steroids. The oxygenation at the 11 position of such steroids is well known to be able to be effected by microorganisms. (16)

My subprogram that discovers functional groups required a time equivalent to 1250 store instructions to find all the functional groups of PGE2 and 2844 store instructions to find all the functional groups of tetracycline. (The actual times were 0.40 and 0.91 ms, respectively, the standard deviation being less than 1%). Comparative data in the literature were not found. The process is quite direct and both the need and the possibility of improving it appear less than for the canonicalization process.

Support for this work from the National Research Council of Canada is gratefully acknowledged.

Appendix. Steps of the sum algorithm in detail sufficient for programming.

1. Calculate the initial canonical values for all the atoms by copying the first 32 bits of each connection table row and shifting right two bits to eliminate the chirality information.

2. Calculate the second and third canonical values for all the atoms. The ith canonical value of an atom is the sum of the i-1th canonical values of its ligand atoms.

3. Make a list of all pairs of atoms which are equivalent in their third canonical values as well as their first canonical values. Call the list S, for Suspected equivalent pairs.

4. Check through the list S to find the following specific cases: gem dimethyl pairs, gem dihalo pairs, methylenes or oxygens of the 1,3-dioxolane ring of a ketal or acetal of ethylene glycol, pairs of corresponding atoms in the carboalkoxy groups of a malonic ester, ortho or meta carbons of monosubstituted benzenes. (Almost all cases I have encountered are covered by this list.) Remove these cases from the list of suspected equivalences and put them on a new list called C. If S is now empty then go to step 7.

5. Calculate the next set of canonical values for all the atoms. Then examine the list S of suspected equivalent pairs of atoms and check to see if their latest canonical values are still the same. If the canonical values are now different remove the pair from S. If S is not empty then go back to the beginning of this step.

6. Find out whether the ligand atoms of each pair of equivalent atoms on the list S are pairwise equivalent. If not then rank the pair of atoms according to the canonical value of the highest ranking ligand. If there are nonterminal double bonds in the molecule then calculate the stereochemical canonical values using the string of first numbers of each pair in the stereochemical table as initial canonical values. Next calculate the second and third stereochemical canonical values. If all the pairs of S are still equivalent then go to step 7. Otherwise remove the inequivalent pairs from S and continue to calculate stereochemical canonical values and check for inequivalences of pairs in S until the length of S no longer decreases.

7. Sort the rows of the connection table and the stereorelation table according to the precedence determined first by the initial canonical values and in case where the initial canonical values are equal, by the final canonical values. In this process prepare a translation table such that the ith entry of the table is the new number of old atom number i.

8. Renumber the ligand atoms in each row using the table just created.

9. Sort the ligands in each row. If an old number of interchanges are required in the sorting process for the row of

an atom which is chiral in the product, then invert the chirality. In this case the chirality of the atom in the reactant is opposite to that which it has in the product.

10. For all pairs on the lists S and C, find any common neighbors and erase any chirality derived from the product connection table.

11. For all pairs of atoms on S or C see if the following rule is obeyed: If i and j are equivalent and i < j then the first ligand of i must have the same or a lower row number than the first ligand of j. If the first ligand of i is the same as the first ligand of j, i.e. they have a common ligand, then the second ligand of i must have the same or a lower row number than the second ligand of j. If the rule is not obeyed, interchange the renumber atom i as atom j, rename the former atom j as atom i and change the translation table, the connection table and the stereorelation table accordingly. Note: I use a selection sort for sorting the rows of the connection table in step 7. For the number of atoms involved, usually 25-40, the straight selection sort was noticeably faster than the tree selection sorts discussed by Knuth. (17) The optimum sorting procedure in this case is very much an open question. The numbers of the ligands in each row were sorted by the bubble sort (17) because these numbers are usually in order before canonicalization, since they are usually in the same order in the reactant molecule at hand as they are in the already canonicalized product molecule connection table. The most efficient sort for an almost sorted list is the bubble sort.

Abstract

The speed of generating reactants is discussed from the standpoint of improvement in two key routines, the canonicalization of molecular structure and the discovery of functional groups. Algorithms in use for canonicalization of molecular structure representations in a computer are classified into tree algorithms and sum algorithms. A particular sum algorithm is described. The advantages of canonicalization are presented, including the discovery of atoms equivalent under local or global symmetry operations. A method for finding functional groups is also described, the essence of which is the special labelling of the connection table rows of atoms that are central to simple functional groups. The product functional group information is used when examining the reactant(s), so that groups can be copied, where possible, and not rediscovered.

Literature Cited

1. Susseguth, E.H., J. Chem. Doc., 5, 36 (1965). A more rapid procedure using a connection table with much information about each atom, is to be found in the appendix of reference 10.
2. Cahn, R.S., Ingold, C.K. and Prelog, V., Angew. Chem. Informatn. Edn., 5, 385, 511 (1966).
3. International Union of Pure and Applied Chemistry, "Nomenclature of Organic Chemistry", Sections A, B & C, Third Ed., Butterworths, London, 1971.
4. Wipke, W.T. and Dyott, T.M., J. Amer. Chem. Soc., 96, 4825, 4834 (1974).
5. Gelernter, H., Sridharan, N.S., Hart, A.J., Yen, S.C., Fowler, F.W., and Shue, H., Topics Current Chem., 41, 113 (1973).
6. Smith, E.G., "The Wiswesser Line-Formula Chemical Notation", McGraw-Hill, New York, 1968.
7. Morgan, H.L., J. Chem. Documentation, 5, 107 (1965).
8. Gasteiger, J., Gillespie, P., Marquarding, D., Ugi, I., Topics in Current Chemistry, 48, 1 (1974).
9. Esack, A. and Bersohn, M., J. Chem. Soc., Perkin I, 1124 (1975).
10. Bersohn, M. and Esack, A., Chemica Scripta, 6, 122 (1974).
11. Lynch, M.F., Harrison, J.M., Town, W.G. and Ash, J.E., "Computer Handling of Chemical Structure Information", McDonald, London, 1971.
12. Buehler, C.A. and Pearson, D.E., "Survey of Organic Synthesis", John Wiley & Sons, New York, 1970.
13. Harrison, I.T. and Harrison, S., "Compendium of Organic Synthetic Methods", Vols I and II, John Wiley & Sons, New York, 1971, 1974.
14. Bersohn, M. and Esack, A., J. Chem. Soc., Perkin I, 2463 (1974).
15. Bersohn, M. and Esack, A., Chemica Scripta, in press.
16. Fonken, G.S. and Johnson, R.A., "Chemical Oxidations with Microorganisms", Marcel Dekker Inc., New York, 1972.
17. Knuth, D.E., "Sorting and Searching", Addison-Wesley, Reading, Mass., U.S.A., 1973.

7

An Artificial Intelligence System to Model and Guide Chemical Synthesis Planning by Computer: A Proposal

N. S. SRIDHARAN

Rutgers University, Computer Science Dept., New Brunswick, N.J. 08903

One of the central problems in applying computer methods to chemical synthesis planning is Search Guidance. The ability of a chemist to guide an interactive computer in its search for syntheses and the possibilities for self-guidance in Artificial Intelligence programs are both limited by the form and content of the search information that is made available as the exploration proceeds. A system for Search Modelling is proposed in this paper which can augment existing systems for synthesis planning and serve to gather, analyze and amplify the information generated during controlled exploration. The search management model is specified in a simple descriptive form and two example models are included in the paper. The guidance of search using the information gathered is specified by a rule set in the simple syntax of Condition=>Action pairs. A chemist can interactively modify this rule set.

Further, an advanced search model is presented which, by introducing the powerful concept of a Planning Space, allows the search for syntheses to go forward, backward and to leap into the middle under controlled conditions.

INTRODUCTION

Planning chemical synthesis routes for known molecular structures is a rich problem area offering a challenge that is being met in inventive and imaginative ways not only by chemists, but also by computer scientists and mathematicians. I bring to this task the perspective of a specialist in the methods of developing Artificial Intelligence within a computing system, and a persistent concern for using the mechanisable aspects of human knowledge and human problem solving techniques in the medium of a machine. I bring to this task no expertise in organic chemistry or in synthesis. But I do have the benefit of several years of intimate contact with the problem of mechanizing the search for syntheses and with expert chemists, especially Prof. W.F. Fowler, who foresaw such possibilities and who worked with us. My application to this task under the guidance of Prof. Gelernter culminated during 1971 in the running version of SYNCHEM I, the first computer program to perform successful multi-step synthesis explorations automatically without on-line guidance or intervention. This first version of SYNCHEM employed several key ideas and techniques that were discovered by others early in the research on mechanical problem solving. These key ideas will be reviewed below.

Since my main interest lies in the direction of artificial intelligence, I have spent the five years following my work in SYNCHEM in working on the application of artificial intelligence methods to problems in Mass Spectrometry with the Heuristic Dendral project at Stanford University and also with Professor Ivar Ugi at Munich in assimilating his algebraic approach to the representation of reactions. Upon Todd Wipke's invitation to participate in this Symposium I have elected to set forth in a very specific manner my proposals on how I would tackle the task of synthesis planning in the light of my current understanding of the advances that have been made in the methods of artificial intelligence.

Thus, I have three aims in writing this paper:

a) Propose a system to augment existing synthesis search systems by focusing on the issues of search management. To this end the concept of search modelling is introduced and two search models are presented.

b) Clarify the fine distinction between selection of transform by Relevance criteria and by Applicability criteria. Selection by Applicability defines what is usually called the State Space and search in the state space usually grows the synthesis path uniformly from one end only. Selection by Relevance is not used by any existing system which when used yields the powerful Planning Space. The search in a planning space can take "leaps" along the synthesis sequence.

c) Indicate that a combination of search in both the state space and the planning space is possible and that this a function of the search model employed. The second search model described in the paper allows the search for synthesis to go forward, backward and to leap into the middle under controlled conditions.

It is hoped that the advantages of this way of developing a system will include: upgrading the role of the chemist from one of rating, pruning and selecting subgoals or precursors to that of giving injunctions to the system in the form of rules added, removed or modified as the search proceeds a few steps at a time; the introduction of the strategic and judgmental knowledge in the program in a manner that decoupled from the knowledge of reaction chemistry; and the ability to experiment easily with various models of search management that are qualitatively different from each other. The payoffs foreseen are of genuine importance to those of us concerned with the techniques of artificial intelligence and when proven practicable should be exciting and challenging to chemists as well.

It must be stated at the outset that these new ideas on search modelling are an outgrowth of my attempts to develop a system for common sense reasoning about human actions using a psychological theory of act interpretation [Schmidt, 1976; Sridharan, 1975; 1976]. This system has capabilities of selecting and applying transforms associated with act names, structuring them into a plan and reasoning both forward and backward along the plan structure. The strategy of act interpretation is coded in the form of rule sets. The framework adapted for this work is called Meta Description System (MDS) [Srinivasan, 1973; 1976], designed and developed by Srinivasan. I shall explore here in detail the use of this framework in the management of search for chemical synthesis. The connections between the psychological theory and synthesis planning strategies will be left for later exposition.

There are three major conceptual approaches [See Sridharan, 1974 for a review] that have been taken on the issue of a computer mediated design of chemical synthesis plans and they share a central idea among them - that of searching a space of possibilities in a systematic manner using empirical knowledge as appropriate. The very power of these approaches stems from the systematization of search of the space of possibilities.

For those approaching synthesis as fertile grounds for designing and building computer programs that solve difficult intellectual problems [Sridharan, 1971, 1973, 1974; Gelernter, 1973, 1976] the main problems are twofold. First, the acquiring and packaging of knowledge of the reactions in a form suitable for use within a program has to be done carefully and the success of the program depends upon the correctness and extent of the knowledge base. Second, the techniques of conducting search with an incomplete, uncertain and possibly inconsistent knowledge base have to be customized to the task of chemical synthesis.

The interactive problem solving concepts developed [Corey, 1969; Wipke, 1973, 1976] are attractive because of their thoroughness and the chemist user tends to approach the system with hopes that the disciplined exploration of pathways will ensure him that he has not overlooked any reasonable alternative.

The third approach taken to computer methods in chemical synthesis is one of formalizing the set of possible reactions [Ugi, 1976], limiting the use of empirical knowledge to the selection rules for these reactions. This method seeks out novel reaction schemes and can suggest to the chemist routes that could not be found by the other methods. However, a reasonable synthesis program is yet to emerge from this approach. The algebraic characterization of the search space however offers the potential of highly interesting structures to be defined on the search space and might be very successful in the long run.

All of these methods currently utilize a search structure generally called the Heuristic Search Method.

THE HEURISTIC SEARCH METHOD

The Heuristic Search Method [Nilsson, 1971; Newell & Simon, 1972] is a usual first approach to problem solving if the specification of the problem itself is given precisely as a _Goal Situation_, and the solution required is some sequence of _Transforms_ i.e. _Operators_ that can effectively transform the current situation to the goal situation. It is important that the allowable operators are finite and are well-defined. The problem solving procedure involves the use of a _Search Model_ that is used in guiding the search for a solution. The heuristic nature of the procedure arises from the use of _approximate_ methods of evaluating progress towards a solution and of assessing the merit and potential of any partial solution sequence. The user gives up _in principle_ the completeness and optimality of the search process gaining _in fact_ more frequent demonstrations of successful problem solving [Newell & Simon, 1972].

In chemical synthesis the goal state is the target molecular structure and the operators to consider for constructing a solution sequence are molecular reactions. The search for a synthesis plan proceeds _backwards_ from the target molecular structure by considering and applying reactions in the _retro-synthetic_ direction. The process is a _serial_ one and has an "inner loop" that repeatedly asks itself the questions "Should the process stop now?" and "What next?". The process can be successful, interesting and powerful depending on the use of a proper Search Model to answer these questions.

The answer to the question "What next?" comes in two parts: the choice of a molecular structure that can be set up as a subgoal to complete the synthesis, and the choice of the operator to try on that subgoal. The information available to the process in arriving at the answers is of two kinds, both of which must be included in the search model. The first kind comprises the catalog of starting materials, the library of reactions, tests of applicability of the reactions, the a priori merit rating for the goodness (yield, specificity etc.) of the reaction and includes in general information that is made available to the system prior to the actual statement of the problem to solve. The nature and extent of this prior information determines the structure of the search possibilities. The other kind includes information that becomes available as the search proceeds and constitutes a rich body of information that is specific to the problem at hand. Only limited examples of the use of the second kind of information can be shown in current programs, perhaps the clearest one is the placement of protection reactions for sensitive functional groups. The manner in which such problem specific information is collected, organized and used to guide search determines the character of the actual search performed for a given problem.

THE PROBLEM SOLVING GRAPH AS A MINIMAL SEARCH MODEL.

The Problem Solving Graph (PSG) [Gelernter, 1962] is a graphical model of the dynamics of the search that is conducted on a given problem, and consists of a root node representing the target molecule linked to a cascade of descendants which are one-level precursors to those at a level higher in the PSG. The Heuristic Evaluation function assigns numerical weights (or in some cases elements of an ad hoc scale of discrete values) to each node in the PSG. Typically, the node evaluation is based not only on properties of the situation designated by the node and

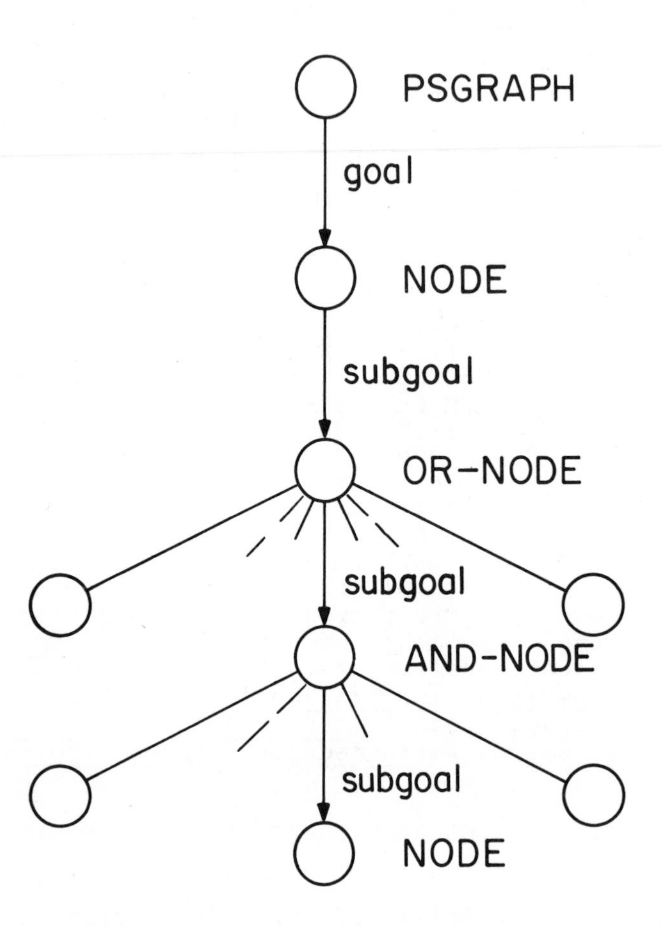

Figure 1. And-or problem solving graph model

the operator applied to generate the node, but also on the entire path from the goal node to the node in question. The PSG model in combination with the heuristic value assignments provides the search mechanism ready answers to the "What next?" question.

The structure of the PSG for the chemical synthesis problem, shown in Figure 1, consists of AND-nodes when operators generate multiple precursors all of which need to be made available for the reaction to be successfully executed, and OR-nodes designating the choice available among several operators applicable to a node. The selection criteria that are well entrenched in the Theory of Heuristic Search [Slagle, 1971] for handling such AND-OR problem solving graphs are:

> At an OR-node select the most promising subgoal;
>
> At an AND-node select subgoals starting from the most likely to fail to the least likely to fail.

INFORMATION-GATHERING AS COMPLEMENTARY ACTIVITY TO HEURISTIC SEARCH.

The planning activity involves a variety of decisions that require information not customarily included in the initial specifications of a transform. Such decisions involving <u>reasoning about the search process</u> beyond the information provided by the PSG model call for a more elaborate search model and is considered in this section.

The PSG model may be viewed as a collection of ready answers to the following set of questions:

a) Does a node have subgoals? How many? What are they?

b) What is the status of a node? Was a successful path found? Was it tried and failed on all paths? Are there descendant subgoals still open?

c) Does the situation (i.e. molecule) represented in

this node occur elsewhere in the PSG? Is the situation circular i.e. calling for synthesizing X to synthesize X higher up?

d) At an OR-node what is the best subgoal to take up? At an AND-node what is the precursor that should be tackled first?

Now consider the following set of questions that go beyond the information maintained by the PSG model:

a) When there is an operator whose relevance to synthesizing a molecule is clear, but the applicability conditions are not satisfied, under what set of circumstances should the unsatisfied preconditions be made into subgoals? Under what conditions should the operator be rejected altogether?

b) For a given molecule what is the most strategic sequence for the introduction of functional groups?

c) When is the sequence of functional group introduction immaterial?

d) Given a subgoal, can the paths explored for another structurally homologous structure be considered valid here?

e) Given a synthesis route (or partial route) involving a protection/unprotection reaction pair, should an attempt be made to derive a revised route not involving protection by resequencing some of the reactions?

Answering these questions calls for maintaining a richer and better organized base of information about the search paths than that provided by the PSG and its heuristic value assignments to the nodes.

As an alternative to the Heuristic Search process, consider the following two-stage process:

a) Perform some exploration in the search space (be it the State Space of the Heuristic Search described above, or the Planning Space to be discussed below).

b) Gather the search information, analyze, amplify and

use it to guide further exploration.

A great deal of flexibility and investigative power comes to us if we separate the total system into two components giving explicit charge of the exploration by search to one, and the analysis and assimilation of the search information to the other. We gain conceptual clarity in thinking about the rules for search guidance and set about designing novel Search Models with a new ease and vigor. I will describe briefly the information gathering system which has been developed at Rutgers and show by example a novel form of search model in the remaining sections of the paper.

META-DESCRIPTION SYSTEM:
A PARADIGM FOR INFORMATION-GATHERING The structure of a system described in the facility of MDS [Srinivasan, 1973 & 1976; Sridharan, 1975] concepts is radically different from the procedure based systems to which we are accustomed. It is more favorable, therefore, to introduce the system directly by an example.

Table I presents the STRUCTURAL DESCRIPTIONS of the classes of entities that are involved in building the PSG model. The class PSGRAPH designates the Problem Solving Graph whose ELEMENTS are NODES. There are two important classes that help structure collections of NODES into OR-NODES and AND-NODES. The latter three classes have MERIT and STATUS relations associated with them, shown in the Table relating these nodes to INTEGER values for MERIT, and a class called STATUS for the STATUS relation. There are three values of STATUS defined as constants viz., OPEN, FAILED and SUCCEEDED. The PSGRAPH has a GOAL which is a NODE and a collection of open nodes and failed nodes. The TRIALNODE designates the node to take up as subgoal when the search process is set to explore the space again. The PSGRAPH has a relation STATUS which is intended to indicate the conditions under which the process should terminate.

NODES are related to the OR-NODEs via the SUBGOAL relation, the OR-NODEs indicate CHOICEs of AND-NODEs and the AND-NODES in turn INCLUDE any number of NODES

```
      (* TABLE I *)

        (* CONSTANTS OF THE DOMAIN PSG *)

(CONSTANTS (YESNO (YES NO)))
(CONSTANTS (STATUS (SUCCEEDED FAILED OPEN)))


        (* STRUCTURAL DESCRIPTIONS *)

(TDN: [PSGRAPH (element NODE elementof)
               (goal NODE goalof)
               (opennodes NODE opennodeof)
               (failednodes NODE failednodeof)
               (status STATUS statusof)
               (trynode NODE|AND-NODE|OR-NODE
                        trynodeof)])


(TDN: [NODE (subgoal OR-NODE subgoalof)
            (subnode NODE subnodeof)
            (descendant NODE descendantof)
            (status STATUS)
            (situation MOLECULE)
            (merit INTEGER meritof)
            (repeated NODE repeatedby)
            (circular NODE)])


(TDN: [OR-NODE (subgoal AND-NODE)
               (status STATUS)
               (merit INTEGER)])


(TDN: [AND-NODE (subgoal NODE)
                (status STATUS)
                (merit INTEGER)])


(TDN: [MOLECULE (structure CHEMICAL-GRAPH)
                (available YESNO)])

               (* SENSE DEFINITIONS *)

(QSCC: [((NODE N) | (X elem N) (N status OPEN))
        PSGRAPH
        opennodes])

(* Flags specifiable on the relations
   have been left out for simplicity *)
```

Table I. Continued

```
(QSCC: [((NODE N) | (P elem N) (N status FAILED))
        PSGRAPH
        failednodes])
(QSCC: [((STATUS S) | (X (goal status) S))
        PSGRAPH
        status])
(QSCC: [((STATUS S) |
          ((X (subgoal status) SUCCEEDED)  =>
                  (X status SUCCEEDED))
          (((NOT [X (subgoal status) FAILED])
            AND
            (NOT [X status SUCCEEDED])) =>
                  (X status OPEN))
          (((ALL NODE N)(X subgroup N)(N status FAILED)) = >
                  (X status FAILED)))
        OR-NODE
        status])
(QSCC: [((STATUS S) |
          ((X (subgoal status) FAILED) =>
                  (X status FAILED))
          (((X (subgoal status) OPEN)
            (NOT [X status FAILED])) =>
                  (X status OPEN))
          (((NOT [X status FAILED])
            (NOT [X status OPEN])) =>
                  (X status SUCCEEDED)))
        AND-NODE
        status])
(QSCC: [((NODE A) | (A (subgoal subgoal subgoal) X))
        NODE
        subgoalof])
(QSCC: [((NODE A) |
          (X subgoalof A)
          OR
          (X (descendantof subgoalof) A))
        NODE
        descendantof])
(QSCC: [((STATUS S) |
          ((X (situation available) YES) =>
                  (X status SUCCEEDED))
          ((X (subgoal status) SUCCEEDED) =>
                  (X status SUCCEEDED))
          ((X circular Y) => (X status FAILED)))
        NODE
        status])
(QSCC: [((NODE R) | (X (situation situationof) R))
        NODE
```

Table I. Continued

```
        repeated])
(QSCC: [((NODE C) | (X repeated C) (X descendantof C))
        NODE
        circular])
(QSCC: [((INTEGER I) |
          (X (subgoal merit) I)
         (NOT [X (subgoal merit >=) I]))
        OR-NODE
        merit])
(QSCC: [((INTEGER I) |
          (X (subgoal merit) I)
         (NOT [I (>= meritof subgoalof) X]))
        AND-NODE
        merit])


        (* PRODUCTION RULES GUIDING SEARCH *)

[INITIALIZE (IT (PSGRAPH P))
            (INPUT (P goal G))
            (IR (P goal G))
            (IR (P opennodes G))]

(G status SUCCEEDED) =>
        (OUTPUT G)(HALT)

(G status FAILED) =>
        (OUTPUT G)(HALT)

(G element X)(X circular Y) =>
        (ASSERT (X status FAILED))

(NOT [G trynode X]) =>
        (ASSERT (G trynode (G goal)))

(G trynode X)(NOT [X subgoal Y]) =>
        (ASSERT (G trynode NIL))(SPROUT X)(EVALUATE X)

(G trynode X)(X subgoal Y)(X (merit meritof) Y) =>
        (ASSERT (G trynode Y))

        (* End of Table I *)
```

thus clearly exhibiting the AND/OR nature of the graph.

Let us turn our attention briefly to the SENSE DEFINITIONS which are specifications of the Logical conditions that are to be met for asserting the various relations and are at the same time specifications of the computations to be performed if the system is given the responsibility to fill in values for certain relations. Simple definitions are given for the OPENNODES and FAILEDNODES relations of the PSGRAPH class. The OPENNODES are the set of all NODEs N which are ELEMENTS of X (denoting the PSGRAPH) whose STATUS is OPEN. The STATUS of the PSGRAPH is defined to be the status of the goal node of the PSGRAPH written
[(STATUS S) | (X (goal status) S)].
The nature of the AND-NODES and the OR-NODES is clearly spelled out in the definitions for the STATUS relations on these classes. The definition for the status of the AND-NODE may be paraphrased into English as follows:

a) If X includes a node whose status is FAILED then the status of X is also FAILED;

b) If X status is not FAILED and X includes an OPEN node then the status of X is OPEN;

finally, c) If X is neither OPEN nor FAILED then it must be SUCCEEDED.

The information displayed in Table I is the description the user provides to the Information-gathering system as the specification of the Search Model. The system accepts the Structural Descriptions and sets up Data Structures and access functions for each of the relations. The Sense Definitions are analyzed to compile a Network of Information Flow [Sridharan, 1976] that prescribes the data flow paths when a new piece of information is made available to this system. This much could be termed the "compile-time" activity of the system.

```
                    (* TABLE II *)

                  (* STRUCTURAL DESCRIPTIONS *)

(TDN: [SEARCHGRAPH
          (* This is the search graph of FLEXI)
                  (elements NODE)
                  (goal NODE)
                  (trynode RNODE|SNODE|DNODE|FNODE)])

(TDN: [SNODE
            (* State Space Structure
             (Backward Search))
            (node NODE snode)
            (subgoal OR-NODE)
            (merit MERIT)
            (status STATUS)
            (subgoal NODE)
            (circular SNODE)
            (features FEATURE)
            (descendant SNODE)])

(TDN: [RNODE (* Planning Space Structure)
            (node NODE rnode)
            (redgoal RNODE)
            (difgoal DNODE)
            (merit INTEGER meritof)
            (status STATUS)
            (relevantfeatures FEATURE)])

(TDN: [DNODE (* State Space Structure
               (Forward Search))
            (node NODE dnode)
            (tonode NODE)
            (fromnode NODE)
            (differences FEATURE)
            (merit MERIT)
            (status STATUS)])

(TDN: [FNODE (* Non-Goal-Directed
               Forward Search)
            (node NODE fnode)
```

Table II. Continued

```
            (transform TRANSFORM appliedto)
            (reactivegroup FEATURE)
            (canproduce FNODE canbeproducedfrom)
            (merit MERIT meritof)
            (status STATUS)])


(TDN: [TRANSFORM
                (* One is needed for
                   every growth)
                (appliedto FNODE transform)
                (product FNODE)
                (reagents MOLECULE)])


                (* SAMPLE PRODUCTION RULES *)


(ADDED (SNODE S)) =>
        (ASSERT ((S node) fnode NIL))
                (* No forward exploration if S was
                   given as a retrosynthetic goal)


(ADD (FNODE D))=>
        (FILLIN (INSTANTIATE
          (DNODE (fromnode &(N node))
                 (tonode &(N (canbeproducedfrom
                            node dnode tonode)))))

        (* If N was generated by forward
           exploration try asserting a DNODE
           subgoal if the information needed is there)


                (* SENSE DEFINITIONS *)

(QSCC: [((DNODE D) | (D tonode (X goalnode)))
        RNODE
        difgoal])

(QSCC: [((DNODE D) |
          ((X (node status) SUCCEEDED) =>
                (X status SUCCEEDED))
          (((X (difgoal status) SUCCEEDED)
            (X (redgoal status) SUCCEEDED)) =>
                  (X status SUCCEEDED))
          (((X (difgoal status) FAILED)
```

```
OR (X (redgoal status) FAILED)) =>
    (X status FAILED)))
     RNODE
     status])


(QSCC: [((NODE Y) | (X (difnodeof goalnode) Y))
        DNODE
        tonode])

(QSCC: [((STATUS S) |
          ((X (fromnode canproduce* node) (X tonode))
           =>
           (X status SUCCEEDED))
          ((X (goalnode goalnode descendant node)
            (X fromnode)) =>
                  (X status SUCCEEDED)))
        DNODE
        status])

(QSCC: [((STATUS S) |
          ((X (node status) SUCCEEDED)
            => (X status SUCCEEDED))
          ((X (subgoal status) SUCCEEDED)
            => (X status SUCCEEDED)))
        SNODE
        status])

(QSCC: [((STATUS S) |
          (((X (rnode status) SUCCEEDED) OR
            (X (snode status) SUCCEEDED) OR
            (X (situation available) YES)) =>
                  (X status SUCCEEDED)))
         NODE
         status])
```

The planning and problem solving is initiated and controlled by a set of rules that we shall examine presently. The initialization is straightforward and involves creating an instance of PSGRAPH and filling its goal node. This indicates that it is the specification of the goal that triggers the process. The specification of the goal node involves submitting the structure of the molecule to be synthesized.

Turning our attention away from the Rule set that controls the problem solving process, let us consider the information gathering activity caused by the addition of a subgoal node for some operator explored by the search component of the system. This may be specified as a conjunction of precursor molecules which are INCLUDED in an AND-NODE. Consider the action taken when one of the precursors is asserted as a SUBGOAL of the GOAL node. The check of the condition for the (NODE subgoalof NODE) relation indicates that this AND-NODE needs to be introduced as a choice in the OR-NODE pointed to by the GOAL node [this is indicated by the expression (A (subgoal choice includes) X)]. The consequent assertion of the CHOICE relation flows along its data flow path to the STATUS relation of the OR-NODE in question. It is appropriate to point out here that this data flow link was "compiled" when the definition of STATUS was scanned and it was established then that any additions/changes to the CHOICE of an OR-NODE was to take effect in turn on the STATUS relation. A verbal description such as this one cannot describe all the data flow that takes place but hopefully the above explanation conveys the concept of the data flow and consequent "information-gathering" proceeding as per the structural and sense definitions of the Search Model given by the user for this problem domain.

At this point, if the reader will grant that as the results of the exploration conducted by the search component gets fed to the Information-Gathering component the requisite search model will be created or updated as appropriate, we can turn our attention back again to the Search Guidance Rules.

The Rules are written in what is known in the computer science lingo as the "Production Rule Form" [Davis & King, 1975].

The production system takes a sequence of conditional action specifications of the form:

RULE: (Condition to check)=>(Action to take)

In applying one of these rules the left-hand side of the Rule is first tested against the current state of the model and if the test is satisfied the actions of the right side of the rule are performed. There are a variety of rule sequencing methods conceivable, but we shall use only the simplest of them here. The control starts from the beginning of the rule sequence and tries each rule and cycles back to the first rule after the last rule is tried. The execution of a (HALT) in some action component of a rule terminates the entire process.

The rule set for the PSG model is very simple. Initially, the status of the PSGRAPH is examined to see if the process should terminate. The rules written here specify that if the status of the PSGRAPH is FAILED or SUCCEEDED then the graph is output and the process halts. Otherwise, If there is a node marked as a possible node to sprout, the goal node is picked first. On the other hand, the presence of a trynode which has no subgoals indicates that the selection is completed and the action to take is to SPROUT the trynode and EVALUATE it. The responsibility of SPROUT is to choose a transformation, test its applicatbility and give the set of precursors to Modelling System. The Modelling System then updates the model and posts the tree hierarchy, the circularity and status relations. The evaluation also submits its merit rating of the nodes to the Modelling System which in turn, reassigns the merits of the affected AND-NODES and OR-NODES. The implications of the new information so entered are followed by the Modelling System and the information needed by the rule set is provided in a ready form. The control repeatedly flows in a SELECT-SPROUT-EVALUATE loop until halted.

The rule set is flexible and can be changed if a specific synthesis problem suggests a different form of control. It is not difficult to enter syntactic guidance based on the model graph using the distance of a node from the goal, the number of conjuncts at an

AND-NODE etc.. It is also possible to guide the search based on chemical information, for example, to disregard subgoals involving a seven-membered heteratomic ring. It is conceivable that when the system runs interactively the guidance will be changed and experimented with as the search proceeds.

SEARCH IN A PLANNING SPACE

The structure of the search space is determined not only by the collection of transforms available to the system but also by the rules for selecting the transforms to be tried for any subgoal. There are two basic ways for selecting transforms.

a) Selection by Applicability. If the transforms are selected because they guarantee that the target molecular structure will be produced upon their application, the required precursors become subgoals. This is the method of selecting transforms by their applicability in the retrosynthetic direction. The synthesis sequence grows a step at a time ensuring that the target molecular structure will be a product of the reaction sequence developed so far, once the precursors are made available. The space of possibilities determined by this criterion of applicability is termed the STATE SPACE.

b) Selection by Relevance. In some cases a transform is selected because it produces a molecule only similar to the target molecular structure but not exactly the same. When such a transform is applied to a target structure T, it may synthesize some structure S that is similar to T, from a set of precursors P. This breaks the original problem of synthesis into two subproblems,

i) Synthesize P. S(P)
ii) Transform S==>T. TR(S,T)

The transform selected specifies a reaction that converts P to S, and this transform will, in general, constitute an intermediate step in the synthesis sequence. The space of possibilities determined by the criterion of relevance is termed the PLANNING SPACE and in several situations the search toward a

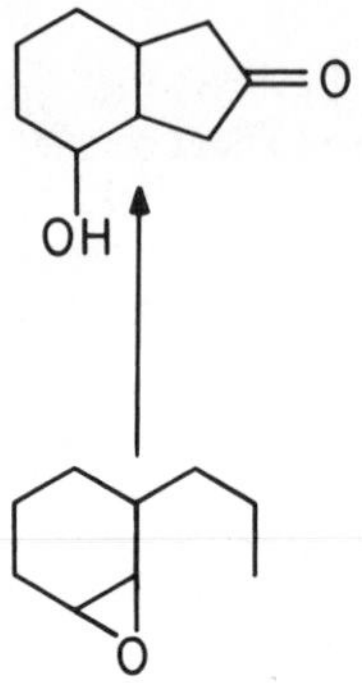

OH

O

Scheme I. Reaction used in transform

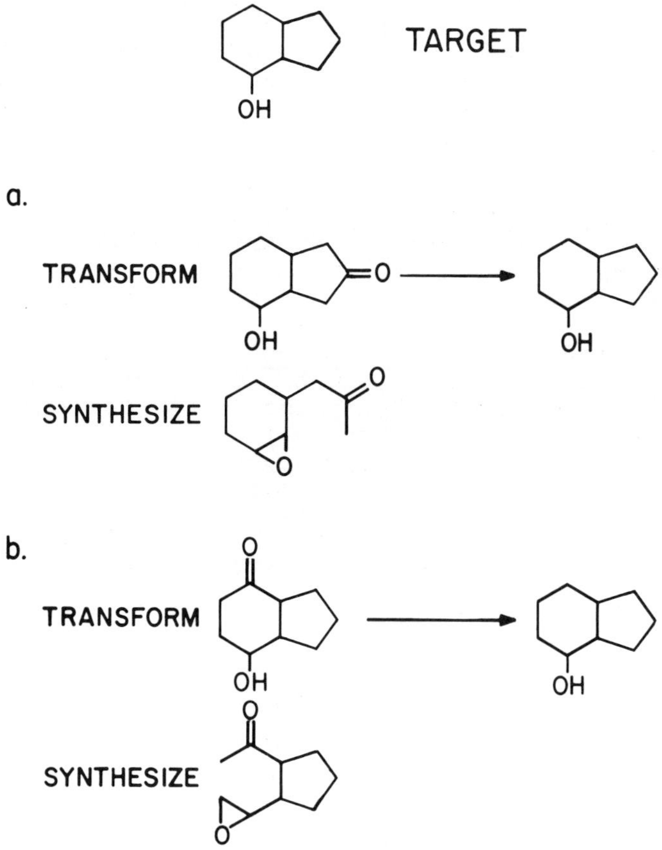

TARGET

OH

a.

TRANSFORM

SYNTHESIZE

b.

TRANSFORM

SYNTHESIZE

Scheme II. Planning space maneuvers

solution can be briefer in this space than in the State Space. This definition of a Planning Space is a variation of the concept introduced in the General Problem Solver system (GPS) [Ernst, 1969].

The distinction between the Selection of transforms by Applicability and by Relevance is an important one when considering the strategies one might employ to search for synthesis sequences. The search in a Planning Space has the characteristic that the search "leaps" into some intermediate point in the synthesis sequence and establishes an "island" and the solution search could then proceed from the island to the target molecule in the forward direction or from the island backward in the retrosynthetic direction toward available molecules. The significance of this ability to leap has been explored in other task areas than synthesis search and has been found to be a powerful tool in converging on solutions rapidly. Its utility for synthesis search remains to be shown and for now can be illustrated only in terms of examples.

The following sketch of an example is offered to illustrate the idea of planning space. The example has not been checked by any chemist and thus its chemical correctness cannot be assured.

Wipke [Wipke, 1976] uses the example of a reaction that synthesizes an alcohol group in 1,4 relation to an electron withdrawing group, say C=O, by the opening of epoxide by a stabilized ion. Consider the target structure shown in Scheme I.

The criterion of applicability would require that the target structure contain the -C(OH)-C-CO- substructure and would not be applicable in the State Space search. In search conducted in the Planning Space, if the transform indicated is considered relevant to the target structure (e.g., if the presence of the alcohol and an unsubstituted 1,4 carbon is sufficient) then this transform may be used. The original synthesis problem is replaced with two subproblems shown in Scheme II.

The above example could be successfully completed by working forward reducing the difference between S and T, and working backward in synthesizing P.

COMBINING SEARCH IN PLANNING AND STATE SPACE

The search in a planning space should be conducted by taking repeatedly the synthesize S type problem for each P splitting it each time into a 'Synthesize' and a 'Transform' type problem, deferring all the Transform TR type problems till some available molecular structure is reached along a path. This will generate a Skeletal Plan where many of the intermediate steps are lacking in detail but each one is given as a TR type problem. If an evaluation function can designed for these skeletal plans much useless search can be avoided by using the planning space.

The search in the state space is conservative and takes small steps attempting to make steady progress towards completing a set of solutions. This can cause either aimless wandering because small changes in the merit values assigned to subgoals cause no significant shifts of attention or because the changes in the merit values cause large abrupt changes in behaviour. This has been given the graphic name of the Mesa Phenomenon by Minsky [Minsky, 1963]. The planning space structures the solution sequences quite differently and causes a more goal-directed search to proceed. The method of transform selection allows the program to "leap" into the solution sequence and decide upon one of the intermediate reactions and permits the solution to grow in both directions. It appears that a judicious combination of both the State Space and Planning Space search methods might be able to overcome some of the difficulties found in each of the methods [Amarel, 1969] With these considerations in view the next section introduces a framework in which to combine the two spaces, letting the chemist user supply the search guidance rules customized to the particular problem at hand.

FLEXI: A FLEXIBLE ADVANCED SEARCH MODEL

Given a collection of nodes designating molecules there are the following types of tasks one can generate (Table II):
a) For a given node whose molecule is not yet synthesized, develop a synthesis by working backwards in the state space using applicable transforms.
b) For a given node whose molecule is not yet synthesized, develop a synthesis by problem reduction using relevant transforms.
c) For a given ordered pair of nodes develop a synthesis route that transforms one molecule to the other.
d) For a given node execute a brief non-goal-directed exploration forwards using reactions in their conventional directions.

A search model is introduced here, called FLEXI, in which four types of structures are used to symbolize the above four categories of tasks and these four nodes form the BUILDING BLOCKS of the search management model. Figure 2 shows that a NODE designates a molecule and individually can be set up as an SNODE for searching retrosynthetic sequence, or as an RNODE for search for a planning route. The FNODE is used to set up the node for forward exploration without a goal guidance.

The sprouting of an SNODE generates a piece of the familiar AND/OR problem solving graph and the status relations on SNODE, OR-NODE and AND-NODE are posted similar to that given earlier.

The sprouting of an RNODE R1, see Figure 3, constitutes a step in the planning space and generates two tasks by problem reduction - an RNODE R2 and a DNODE D1. R2 calls for the synthesis of a molecule by further problem reduction and the DNODE sets up a problem of transforming one molecule into another. The transform used in the sprouting of R1 is used to establish that it canproduce the fromnode N3 of the DNODE from the node N2 of R2. This is exhibited in Figure 3.

The new task set up in the RNODE could of course be restructured as an SNODE task by some rule in the production system. The RNODE will be considered successful as soon as the NODE connected to it has

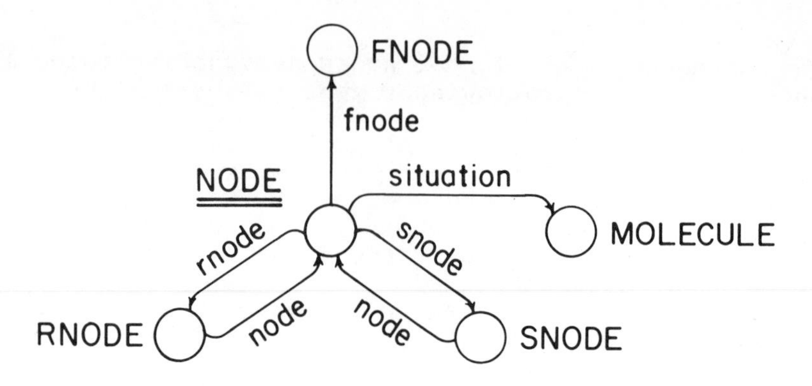

Figure 2. NODE building block

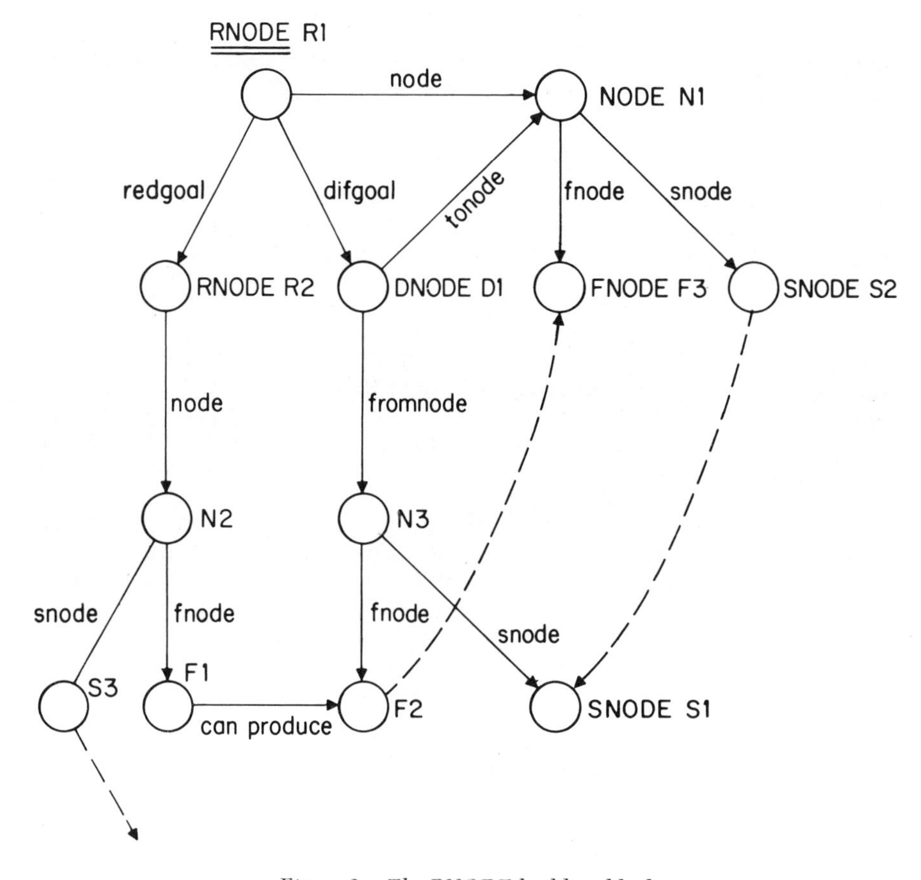

Figure 3. The RNODE building block

SUCCEEDED, whether by its RNODE or the SNODE tasks or even by its designating a molecule which is available in the Catalog of Starting Compounds.

Under certain circumstances a synthesis for the fromnode N3 of the DNODE could be attempted independently and its success will disallow the RNODE R2 from further consideration during search. In that case, the success of the DNODE is sufficient to guarantee the success of RNODE R1 and thereby of NODE N1.

The DNODE can succeed either when a path is found by working forwards from FNODE F2 to F3 or by working backwards from SNODE S2 to S1.

Figure 4 illustrates the canproduce relation among pairs of FNODES. Working forwards from F1 if a molecule M2 results from a transformation involving M1 then the corresponding FNODE F1 can produce F2.

The structures exhibited here are only the building blocks of the search model. By suitably controlling and guiding the exploration, the search can take on great variety traversing the state space forwards or backwards or traversing the planning space. The rule set can be made to contain broad injunctions such as "Do not explore a node in the forward direction if it was created by instantiating a SNODE, i.e. a node to proceed in the retrosynthetic direction" or "When you add an FNODE immediately instantiate a revised DNODE type task" as shown in Figure 5. Of course, these two rules are used here only as examples and in given situations one might wish to advise the system otherwise. The essential point is that a flexible form of search guidance specification is available and can be used to bring to bear on a given problem a wide variety of hints, suggestions and advice that would be difficult in a standard Heuristic Search program.

Overcoming some difficulties of Heuristic Search.

One cause of the Mesa Phenomenon in the case of chemical synthesis is the use of functional group substitution reactions while working in the

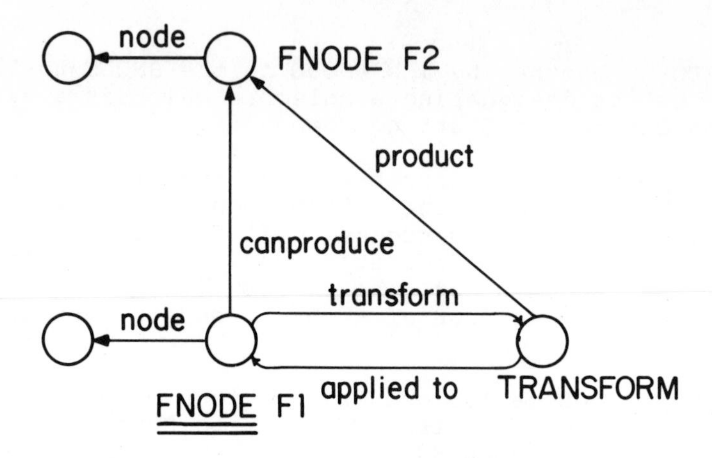

Figure 4. FNODE building block

```
(ADDED (SNODE S)) =>
          (ASSERT ((S node) fnode NIL))
          (* No forward exploration if S was
             given as a retrosynthetic goal)

(ADDED (FNODE D)) =>
          (FILLIN (INSTANTIATE
              (DNODE (fromnode &(N node))
                     (tonode &(N (canbeproducedfrom
                                  node dnode tonode))))))

          (* If N was generated by forward exploration
             try asserting a DNODE subgoal if the
             information needed is there)
```

Figure 5. Sample guidance rules

retrosynthetic direction. If a molecule containing several functional groups is selected for sprouting then performing functional group substitutions often yields

precursors that have nearly the same merit as the target molecule. The presence of several functional groups only aggravates the situation. Within FLEXI this class of reactions could be used mainly for transforming a molecule to another when they are structurally similar, i.e. a DNODE type task which might be carried out most favorably using functional group substitutions. Thus, avoiding the use of these reactions in the retrosynthetic direction should prevent the problem. As further specific problems are isolated and solved, the framework of FLEXI may help us to submit the proper rules of search guidance to the system.

CONCLUSION

The Heuristic Search method is basically very simple. It involves a serial processor working backwards that selects subgoals and transforms by asking "What next?". A good implementation of the heuristic method includes a SEARCH MODEL which is a <u>symbolic representation</u> of the progress of search. The Problem Solving Graph that is commonly used to guide search is a useful but limited search model.

<u>Symbolization is the key to Reasoning</u> and the computer can reason only about things it can handle symbolically. Furthermore, the richness of the search model contributes to the conduct of an intelligent search.

In this paper a search modelling system is described by examples. This system allows the user to <u>describe</u> rather than to program the search model and to associate constraints that govern the growth of the model. The system provides a Rule Language based on the user described search model and the user may then <u>prescribe</u> rules in the form of Production Rules. The rules the user writes can be general, being valid over a wide variety of task specifications, or can be bits of advice and hints about a given problem. The paper concludes by showing how some of the standard difficulties with heuristic search such as getting locked into a plateau can be overcome by suitable techniques of search modelling. The modelling ideas shown here can be used to control and specify protection reactions and for sequencing functional group introduction. It is also possible to carry out different styles of exploration varying the emphasis on the directionality and space in which the search is conducted. The applicability of the model to complement a heuristic search process is now made clear, however its actual use in chemical synthesis planning awaits the participation of a chemist collaborator!

ACKNOWLEDGMENT

I wish to thank Prof. Srinivasan for his stimulating discussions of his modelling ideas represented in MDS without which my understanding of MDS would be meager. A subset of MDS adequate to deal with PSG and FLEXI is being programmed and made available but so far no synthesis problems have been tried out in these frameworks. The present system is implemented in the language FUZZY [LeFaivre, 1974]. I look forward to the continued cooperation of Prof. LeFaivre in the future and his help in the past is hereby gratefully acknowledged. I wish to thank Prof. Saul Amarel for his expert advice upon reading this paper.

REFERENCES

1. Charles Schmidt [1976]
Understanding Human Action: Recognizing the Plans and Motives of Other Persons, in Cognition and Social Behavior, J. Carroll & J. Payne (editors), Lawrence Earlbaum Press, 1976.

2. N.S. Sridharan [1975]
The Architecture of BELIEVER: A System for Intepreting Human Actions. Technical Report RUCBM-TR-46, Department of Computer Science, Rutgers University, New Brunswick NJ.

3. N.S. Sridharan [1976]
The Architecture of BELIEVER-Part II. The Frame and Focus Problems in AI. Technical Report RUCBM-TR-47, Department of Computer Science, Rutgers University, New Brunswick NJ.

4. Chitoor Srinivasan [1973]
The Architecture of a Coherent Information System. Advanced Papers of the Third International Joint Conference on Artificial Intelligence, Stanford, 1973.

5. Chitoor Srinivasan [1976]
An Introduction to the Meta-Description System. Technical Report SOSAP-TR-18, Department of Computer Science, Rutgers University, New Brunswick NJ.

6. N.S. Sridharan [1974]
A Heuristic Program to Discover Syntheses for Complex Organic Molecules. Proceedings of the IFIP74 (International Federation for Information Processing) Congress, Stockholm, August 1974.

7. N.S. Sridharan [1971]
An Application of Artificial Intelligence to the Discovery of Complex Organic Synthesis. Doctoral Dissertation, Department of Computer Science, SUNY at Stony Brook, 1971.

8. N.S. Sridharan [1973]
Search Strategies for the task of Organic Chemical Synthesis. Advanced Papers of the Third International Joint Conference on Artificial Intelligence, Stanford, 1973.

9. H. Gelernter [1973]
The Discovery of Organic Synthetic Routes by Computer. Topics in Current Chemistry, Volume 41, Springer-Verlag, 1973.

10. Herbert Gelernter [1976]
Empirical Explorations of SYNCHEM, An Application of Artificial Intelligence to the Problem of Computer-Directed Organic Synthesis Discovery. This volume.

11. E.J. Corey & W.T. Wipke [1969]
Computer-assisted Design of Complex Organic Synthesis, Science, Volume 166 (178), 1969.

12. Todd Wipke [1973]
Computer-Assisted Three Dimensional Synthetic Analysis. In Computer Representation and Manipulation of Chemical Information. W.T. Wipke et. al. (editors), John Wiley (New York) 1974.

13. Todd Wipke [1976]

SECS--Simulation and evaluation of Chemical Synthesis: Strategy and Planning. This volume.

14. Ivar Ugi [1976]
The Synthetic Design Program MATSYN as a part of MATCHEM, A System of Computer Programs for the Deductive Solution of Chemical Problems. This volume.

15. Nils Nilsson [1971]
Problem Solving Methods in Artificial Intelligence. McGraw Hill (New York) 1971.

16. Allen Newell & Herbert Simon [1972]
Human Problem Solving. Prentice-Hall (New Jersey) 1972.

17. Herbert Gelernter [1962]
Machine-Generated Problem Solving Graphs. Proceedings of a Symposium on the Mathematical Theory of Automata Polytechnic Institute of Brooklyn Press (New York) 1962.

18. James Slagle [1971]
Artificial Intelligence: The Heuristic Programming Approach. McGraw-Hill, New York, 1971.

19. Randall Davis and Jonathan King [1975]
An Overview of Production Systems, Stanford Artificial Intelligence Laboratory Memo AIM-271, October 1971.

20. George Ernst & Allen Newell [1969]
GPS: A Case Study in Generality and Problem Solving. Academic Press (New York), 1969.

21. Marvin Minsky [1963]
Steps Toward Artificial Intelligence. In Computers and Thought, by Edward Feigenbaum and Julian Feldman (editors), McGraw-Hill (New York) 1963.

22. Saul Amarel [1969]
Problem-Solving and Decision-Making by Computer: An Overview. In Cognition: A Multiple View, Garvin (editor), Spartan Books (Washington) 1970.

23. Richard LeFaivre [1974]
Fuzzy Problem-Solving, Ph.D. Dissertation, Computer Science Department, University of Wisconsin, Madison, 1974.
The Representation of Fuzzy Knowledge, Journal of Cybernetics, volume 4, p57-66, 1974.

8

Computer-Assisted Synthetic Analysis in Drug Research

P. GUND, J. D. ANDOSE, and J. B. RHODES

Merck Sharp & Dohme Research Laboratories, Dept. of Scientific Information, and Corporate Management Information Systems, MSDRL Systems and Programming Dept., Merck & Co., Inc., Rahway, N.J. 07065

It was recognized at Merck some time ago (1) that the analytical and data-handling capabilities of a computer could facilitate organic synthesis design, just as spectroscopic methods have revolutionized structure determination. Indeed, the chemist's approach to synthetic design resembles a computer program (2). As outlined in Figure 1, the chemist begins by defining his synthetic problem; he collects relevant knowledge about chemical reactions (note that he never sequentially searches through all available literature); and he designs a synthesis. If the synthesis fails or if he wants additional possibilities, he iterates (repeats the process) to generate additional syntheses.

If initial attempts are unsatisfactory, the chemist may enlarge his store of relevant knowledge, e.g. by reading the literature or talking to an expert; or he may redefine the problem - i.e., find new keys so that more of his knowledge of reactions becomes relevant. In fact, chemists have been known to search exhaustively for a way to implement a preferred route, only to finally re-analyze their problem and find an entirely different - and ultimately successful - one.

If sufficiently desperate, the chemist may browse (i.e., perform a random search) through the literature for ideas. This method has occasionally succeeded when more rational approaches failed, and might be taken as an indication that our reaction classification and retrieval methods are imperfect.

We may envision two levels of computer support of the chemist's analytical process. One approach - which we may call a reaction retriever - organizes and retrieves relevant reaction information. The other, which we here call a synthesizer, simulates a large part of the synthetic process, as shown by the frame in Figure 1. Both computer approaches require a data base of chemical reactions, and it is conceivable that they could share the same data base. We will return to this point; but first, we should consider whether either method would find use by the practicing chemist.

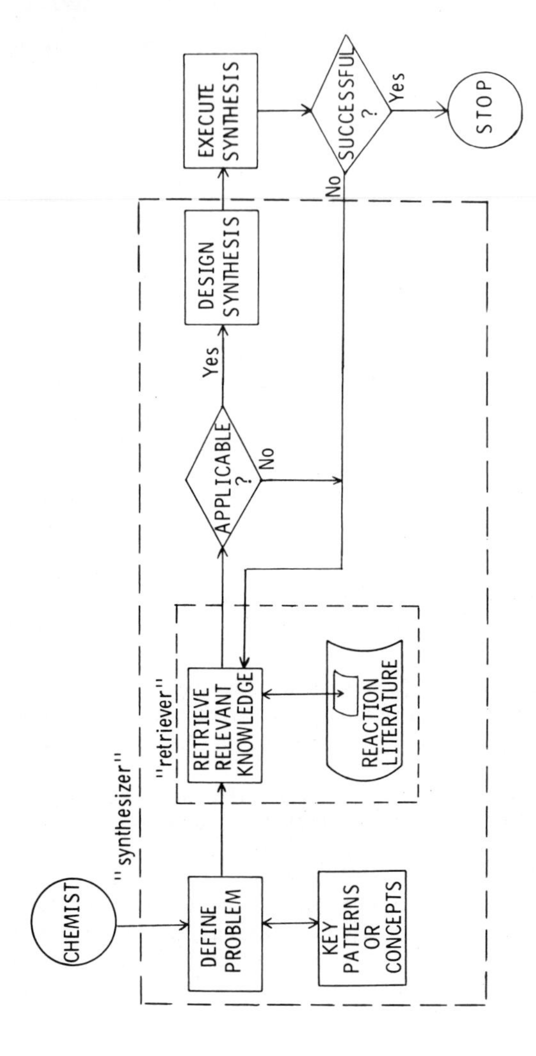

Figure 1. Organic synthesis: Problem analysis

Categories of Pharmaceutical Synthetic Analysis

We identify four types of pharmaceutical synthesis: (1) synthesis of analogs of a "lead" compound, (2) synthesis of natural products, (3) process development, and (4) new reaction discovery (Figure 2). A lead analog program normally attempts

Synthesis Type	Synthesis Class	Computer Aid
Analogs of "Lead"	SM → P	Retriever
	? → P	Synthesizer for Difficult Compounds; Retriever
	SM → ?	Retriever
Natural Product	? → P	Synthesizer; Retriever
	SM → P	Retriever
Process Development	? → P	Synthesizer; Retriever
	SM → P	Retriever
New Reactions	SM → ?	Retriever; Forward Operating Synthesizer

Figure 2. Computer-assisted synthetic analysis types and classes

to find short syntheses of a series of related compounds, often from a single intermediate which can be obtained in large quantities - for example, construction of various side chains starting from penicillanic acid. Occasionally, however, a desired analog must be made by quite different chemistry - for example, preparation of dethiacephalosporins (3). Also occasionally, analogs will be made by applying straightforward chemistry to an available starting material. Thus, if we differentiate three classes of synthetic analyses - (a) starting material and product specified (SM → P); (b) product specified (? → P); and (c) starting material specified (SM → ?), then lead analog syntheses may belong to any of the classes.

When a natural product with interesting biological activity is isolated, synthesis is used to confirm the structure and to obtain sufficient material for biochemical and structure modification studies. Synthetic analysis is generally of the "product specified" (? → P) type, except when a related material is known and available (SM → P synthesis).

For process development, analysis tends to be exhaustive, since the optimal commercial synthesis often is different from the "best" laboratory method. When a cheap related compound is available, the analysis may be of the SM → P type.

Finally, chemists apply new reactions to known compounds in order to generate new drug leads; this requires synthetic analyses of the SM → ? or, occasionally, SM → P types.

A reaction retriever program is applicable to all four types of syntheses. A synthesizer program applies primarily to the ? → P class of problem, although a forward working synthesizer would apply to SM → ? problems. The applicability of these programs by synthesis type is summarized in Figure 2. As a generalization, a synthesizer is most useful for "creative" syntheses, while a reaction retriever may identify optimal conditions after a reaction pathway has been chosen. Therefore both methods should be valuable.

Program Descriptions

A reaction retriever program permits organizing and enlarging the store of reaction knowledge available to each individual. Chemists have long used manual systems for organizing reaction knowledge, such as individual card files, Theilheimer's famous series of volumes, Reactiones Organicae, and recently the Derwent Chemical Reactions Documentation Service. Computer organization of such collections enables retrieval by reactants, products, reaction type, reaction conditions, and/or mechanism (4). Creation of such a computer program is generally considered an information retrieval application (4).

A synthesizer program traditionally begins with a target structure and applies chemical rules to generate and evaluate potential precursors. It offers the capability of fast, exhaustive, unbiased synthetic analyses. As indicated in many of the other contributions to this symposium, this approach is usually considered an artificial intelligence application.

Since we had concluded that both program types were desirable, we wondered if the same reaction data base could serve for both computer approaches, as the flow diagram of Figure 1 suggests. We therefore embarked upon a feasibility study to test this dual use concept.

Proposed Computerized Chemical Reaction Collection

We conceived a system where reactions coded by Merck chemists could serve three purposes - current awareness, reaction retrieval, and synthesizer input (Figure 3). We identified the following system development stages: (I) development of reaction coding sheet; (II) description of reactions on sheets by chemists (continuing); (III) translation to computer readable reaction information by information scientists and typists (continuing); (IV) development of software for computer storage and retrieval of reaction information. In the feasibility study, we actually carried out phases I and II, and performed a limited systems analysis of phases III and IV.

We designed a reaction coding sheet to contain most of the information needed by both reaction retriever and synthesizer programs (Figure 4). Data on reactants, products, intermediates, reagents, conditions, yields, work-up procedures, mechanism, and

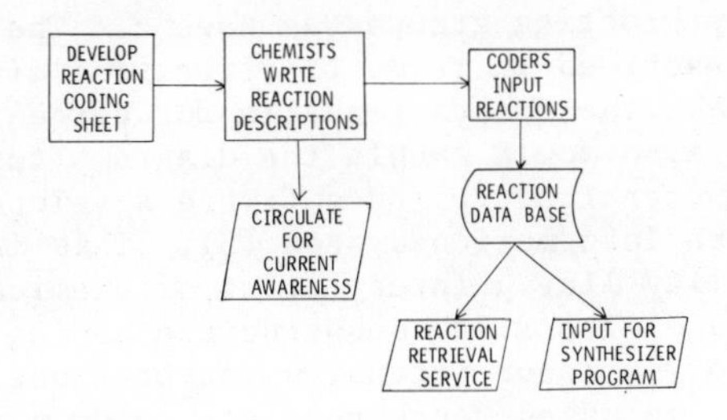

Figure 3. Reaction file development

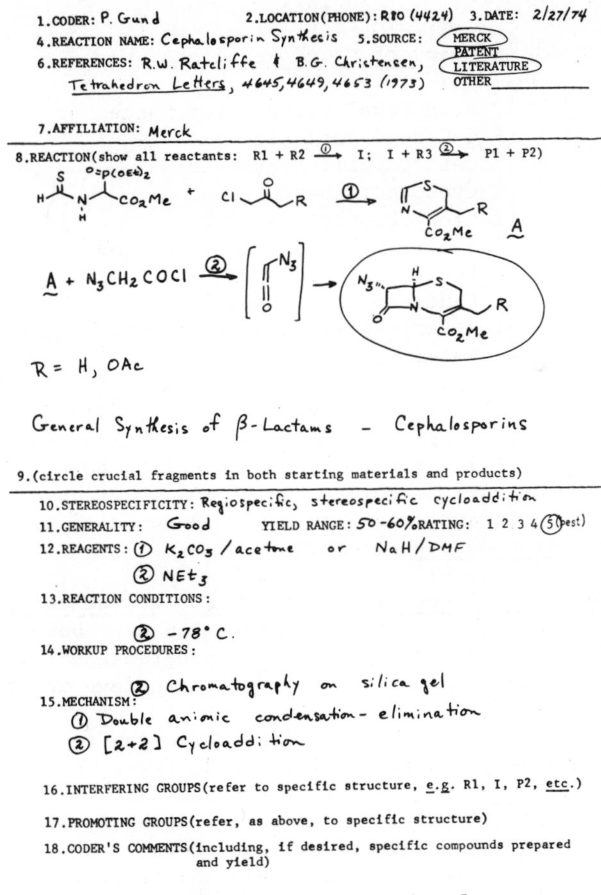

1.CODER: P. Gund 2.LOCATION(PHONE): R80 (4424) 3.DATE: 2/27/74

4.REACTION NAME: Cephalosporin Synthesis 5.SOURCE: MERCK PATENT LITERATURE OTHER________

6.REFERENCES: R.W. Ratcliffe & B.G. Christensen, Tetrahedron Letters, 4645, 4649, 4653 (1973)

7.AFFILIATION: Merck

8.REACTION(show all reactants: R1 + R2 —①→ I; I + R3 —②→ P1 + P2)

R = H, OAc

General Synthesis of β-Lactams – Cephalosporins

9.(circle crucial fragments in both starting materials and products)

10.STEREOSPECIFICITY: Regiospecific, stereospecific cycloaddition

11.GENERALITY: Good YIELD RANGE: 50-60% RATING: 1 2 3 4 5(best)

12.REAGENTS: ① K_2CO_3 / acetone or NaH/DMF
② NEt_3

13.REACTION CONDITIONS:
② -78° C.

14.WORKUP PROCEDURES:
② Chromatography on silica gel

15.MECHANISM:
① Double anionic condensation-elimination
② [2+2] Cycloaddition

16.INTERFERING GROUPS(refer to specific structure, e.g. R1, I, P2, etc.)

17.PROMOTING GROUPS(refer, as above, to specific structure)

18.CODER'S COMMENTS(including, if desired, specific compounds prepared and yield)

Figure 4. Merck reaction coding sheet

interfering and protecting groups was sought. The reaction was expected to be described in terms of structural diagrams; this not only would make the sheets readable for current awareness circulation, but also would enable the diagrams to be coded directly for computer input using software developed for the Merck Chemical Structure Information System (5). This in turn would provide high quality line printer output of chemical structures (see Figure 5), and also allow substructure coding by computer program instead of by labor-intensive, error-prone manual coding.

In phase II, we asked Merck chemists to code reactions in their area of expertise. We received 44 coded sheets from 31 chemists, a gratifying response. However, while these were eminently suitable for current awareness circulation and for a reaction retrieval service, they were generally unsuitable for a synthesizer. Thus, as summarized in Table I, over 50% of the

Table I

Summary of Coded Reactions

18	Functional Group Interchange
9	C-C Bond Formation
7	Heterocyclic Syntheses
2	Cycloadditions
2	Group Protection
2	Functional Group Introduction
2	Functional Group Removal
1	C-Hetero Bond Formation
1	Hetero-Hetero Bond Formation
44	Total

reactions collected were functional group manipulations - probably the class of reaction most commonly performed in the laboratory, but the least useful for generating multistep, nontrivial syntheses. Moreover, few of the reactions were described in sufficient detail to ensure generation of valid pathways by a synthesizer program.

Reactions for a Retriever vs. Reactions for a Synthesizer

Although in principle the same reaction data base could be used for both retriever and synthesizer programs, our feasibility study indicated that this was difficult to achieve in practice, for several reasons.

A synthesizer retrieves reactions according to functionality or substructure of the target molecule, so fewer keys are needed than for a retriever program which requires entry by product, reactants, reaction type, etc. Therefore file organization for one program is not necessarily optimal for the other.

Furthermore, while an incompletely described reaction may be usefully included in a retriever program, it could be

```
CHEM STRUCS SEARCH REPORT MARCH 10, 1976

OW.B.GALL/DR.P.GUND SEARCH 3096 BD SEARCH 14 S TEST
FIND
```

•Q1 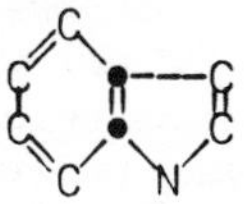 •Q2 Cl—C(a)

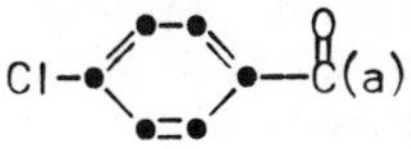

β Q1 AND Q2

*
MERCK CHEMICAL STRUCTURE INFORMATION SYSTEM
*

```
**********************************************************************

L-590,225-00S

C20H18NO4Cl      MOL. WT. 371.823

NS

--
```

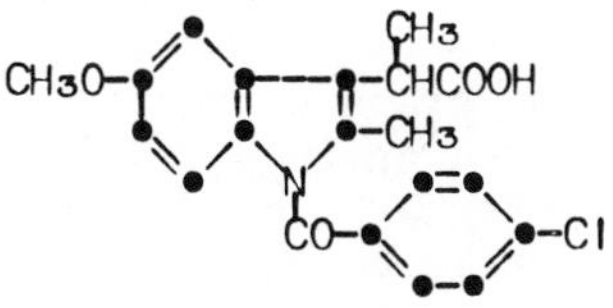

CH3O— CH3 —CHCOOH —CH3 N CO— —Cl

2-/1-P-CHLOROBENZOYL-5-METHOXY-2-METHYL-3-INDOLYL/PROPIONIC ACID

Figure 5. Line printer output of chemical structures

disastrous for a synthesizer. The latter interprets each reaction as a series of chemical instructions for generating precursors. Each reaction description therefore must be tested and debugged to prevent proliferation of numerous improbable synthetic pathways.

Finally, reaction descriptions should ideally be different for these two purposes. While a retriever can handle many very specific reactions, a synthesizer requires generalized reactions for efficiency. Thus, a synthesizer - like the chemist - must initially decide whether, e.g., reducing a ketone to an alcohol is desirable; it is usually in a later step of the analysis that the chemist decides which of the hundreds of reaction conditions for ketone reduction should be tried. A synthesizer program which attempted to apply each of several hundred retro-reduction transforms each time an alcohol appeared in a target molecule, would not be very efficient.

In conclusion, retriever and synthesizer programs are complementary in aiding the synthetic chemist. A synthesizer excels in deriving multistep routes to designated products, while a retriever is superior for choosing specific reaction conditions. The reaction data bases should reflect these differences.

Current Status

We are evaluating the Derwent Chemical Reactions Documentation Service as the basis of a reaction retrieval system. If Derwent succeeds in their objective of coding ca. 20,000 reactions from Theilheimer, Organic Syntheses, etc. - plus coding 4000 new reactions per year - they will create a data base which we could not hope to duplicate ourselves.

For a synthesizer, we are reimplementing the Simulation and Evaluation of Chemical Synthesis (SECS) program on our IBM computer in collaboration with Professor Wipke. Once the program is running at Merck, we will instruct the chemists in program operation, and then ask them to code and test transforms.

While reimplementation of SECS on IBM equipment has proven to be nontrivial, we have made substantial progress. We have successfully interfaced our GT42 graphics terminal with our IBM 370/158 computer under time shared option (TSO), and the major portion of the program is converted.

In the meantime, we have utilized the version of SECS operating on First Data Corporation's time-sharing system to gain familiarity with the program, to aid in debugging our IBM version, and to evaluate the program's capabilities. In one study, performed with the participation of E. Grabowski and R. Czaja, the program demonstrated that it was relatively easy for the chemist to use; that it could produce novel synthetic routes; and that costs were within reason. It proved to be highly desirable to have a knowledgeable chemist guide the program. The analysis also revealed some gaps in the program's chemical knowledge, which clearly represents an area for future development.

Acknowledgements

We wish to thank the Merck chemists who participated in the feasibility study, and our consultant, W. T. Wipke, for his detailed assistance in program conversion.

Literature Cited

1. Sarett, L. H., "Synthetic Organic Chemistry: New Techniques and Targets", presented before the Synthetic Manufacturers Association, June 9, 1964.
2. Corey, E. J., Pure Appl. Chem. (1967), 14, 19.
3. Guthikonda, R. N., Cama, L. D., and Christensen, B. G., J. Amer. Chem. Soc. (1974), 96, 7584.
4. Valls, J., in "Computer Representation and Manipulation of Chemical Information", Wipke, W. T., Heller, S. R., Feldmann, R. J., and Hyde, E., Edits., J. Wiley, New York, 1974, p. 83.
5. Brown, H.D., Costlow, M., Cutler, F. A., Jr., DeMott, A. N., Gall, W. B., Jacobus, D. P., and Miller, C. J., J. Chem. Info. Comp. Sci. (1976), 16, 5.

9

Computer-Assisted Structure Elucidation: Modelling Chemical Reaction Sequences Used in Molecular Structure Problems[1,2]

TOMAS H. VARKONY, RAYMOND E. CARHART, and DENNIS H. SMITH

Departments of Chemistry, Computer Science, and Genetics, Stanford Univ., Stanford, Calif. 94305

Our research in the applications of computer techniques to chemical problems has focused on elucidation of molecular structures of unknown compounds. We have been applying problem solving methods derived from research on artificial intelligence to create a program which emulates certain phases of manual approaches to structure elucidation. This program, called "CONGEN",[3] provides a general mechanism for assembly of chemical atoms and structural fragments inferred from any of a variety of sources. Such fragments ("superatoms"[3]) are inferred manually and subsequently supplied to the program. Statements about structural fragments and constraints on the ways in which they may be assembled are input to CONGEN using a graphical language for representation of structures. This language has important ramifications in extensions to CONGEN, as we outline subsequently.

There are, however, several other important phases of structure elucidation, all of which are amenable to computer-assistance. A representation of major milestones in typical structure elucidation problems is presented in Fig. 1. For the purposes of the subsequent discussion we consider two different categories of structure problems, both of which fit into the scheme of Fig. 1. The first category we view as the general problem of structure elucidation, wherein an unknown compound is isolated and characterized. The second category we term "mechanistic" structure elucidation. Into this category fall synthetic reactions where the precursor, or starting material, is known but the product(s) and the precise reaction pathways are not. Of course there can be some overlap between these categories, and CONGEN handles both in a similar way, but we will discuss them as separate topics. Thus, the "chemical history" collected (Fig. 1) may, in the former case, be actual chemical tests or reactions carried out to characterize the unknown. For mechanistic studies, the history may include a specific reaction exercised on a known structure. Data interpretation and structure assembly will be in the former case actual assembly of inferred fragments, while in the mechanistic case it usually will involve manipulations of, or slight modifications to a known structure.

Until recently, CONGEN performed only structure assembly and some data interpretation and provided the facilities to assist examination of structural possibilities and elimination of inconsistent structures. We are now actively pursuing other elements of Fig. 1. For example, examination of structures and subsequent design of new experiments is an interesting chemical and artificial intelligence problem currently under investigation. Although actual collection of spectroscopic and chemical data is beyond the scope of our current interest insofar as programs for symbolic reasoning are concerned, the element of data interpretation is also of major importance to us.

The use of chemical transforms, or reaction sequences, in structure elucidation (Fig. 1) in both the general and mechanistic senses mentioned previously is the subject of this report. Reactions or sequences of reactions may be carried out on an unknown for several reasons. The reaction may a) test for a specific functional group; b) simplify the problem by decomposing the unknown into smaller, more easily characterizable molecules; c) modify the skeleton or functional groups to define more accurately their respective environments or make the unknown more amenable to analysis (e.g., increase its volatility); or d) unambiguously relate the unknown to a previously characterized compound.

Mechanistic studies, usually involving rearrangements or cyclizations, employ reaction sequences to help characterize reaction pathways and establish relationships among sets of related structures. The multiplicity of pathways open to such processes frequently prevents establishing structures of products without additional collection and examination of data. In such cases, the chemical transform which forms part of the chemical history can be carried out to yield the set of candidate structures

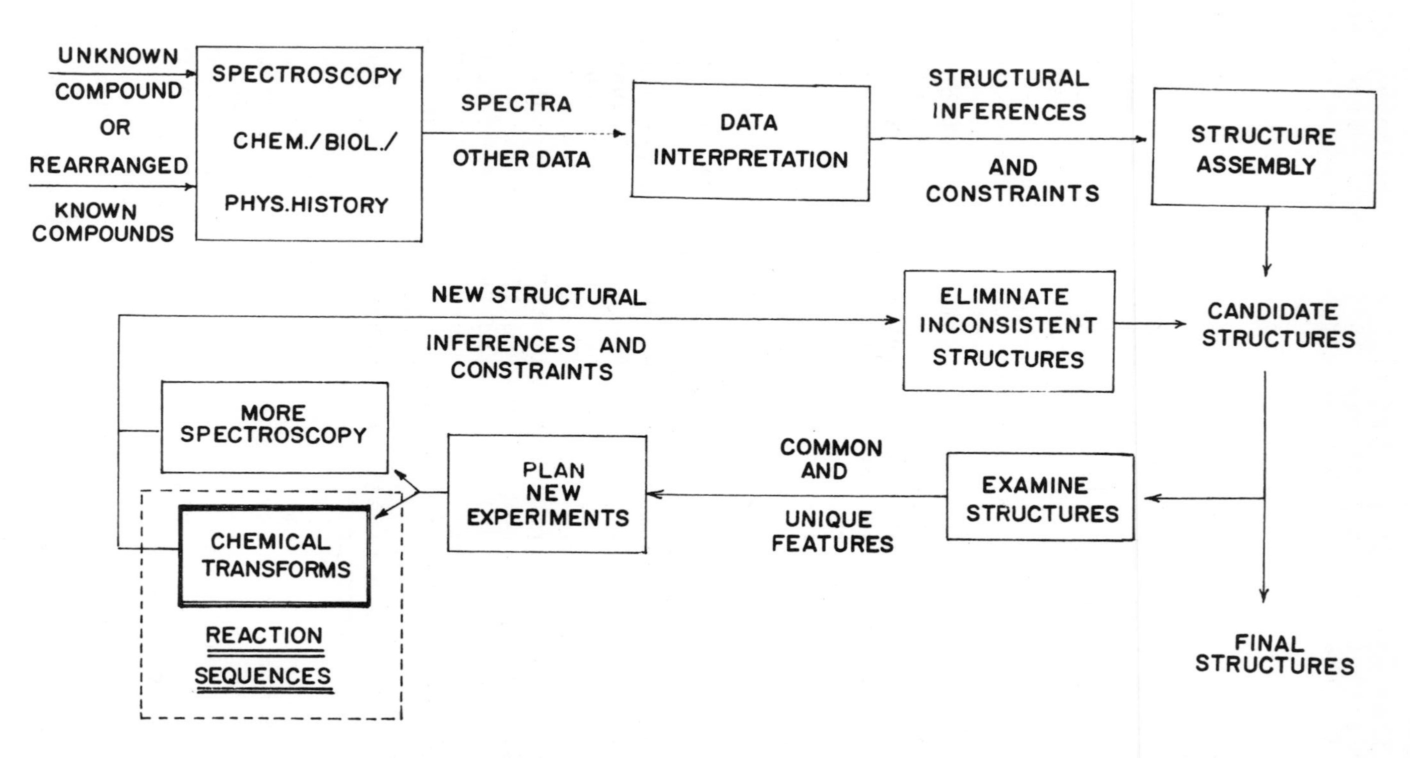

Figure 1. Major milestones in the elucidation of molecular structure. Reaction sequences carried out on a known structure or on candidate structures for an unknown are the topic of this paper.

which may then be treated just as candidate structures for an unknown. However, the application of constraints on the reaction products has different meaning in mechanistic studies as contrasted to general unknowns as we describe in the Methods section.

PURPOSE

Similarities to and Contrasts with Computer-Assisted Synthesis.

As we will illustrate in subsequent sections, our work on chemical reaction sequences has several important similarities to and contrasts with current efforts directed toward computer-assisted synthesis[4,9]. The similarities fall primarily in areas of structure and reaction definition and manipulation. We share common problems of user interaction and interfacing between the outside world and the more rigidly structured domain of the computer program. Reactions and structural constraints on them must be defined and such definitions must be saved to provide a knowledge base which can be called upon by future users. Internal to the programs are common problems of perceiving important molecular features and executing the reaction by appropriate manipulations of the structure representation according to the definition of the reaction. Algorithms common to both problems include ring perception ("cycle finding"), "path finding" to determine connectivity, some form of graph matching to detect given substructures, recognition of symmetry properties of reactants or products, avoidance of or detection and elimination of duplicate structures and representation of chemical as opposed to graph-theoretical concepts of structure, such as aromaticity.

Despite the fact that both our efforts and those of computer-assisted synthesis design involve executing a representation of a chemical reaction in the computer, there are fundamental philosophical and methodological differences, as summarized in Fig. 2 and detailed as follows:

I) A synthesis problem has a specific target molecule. The goal in developing a synthesis is to define precursors which are in some sense simpler. The precursors become the targets for the next level and the procedure is recursive until the terminating condition of sufficiently simple precursors is

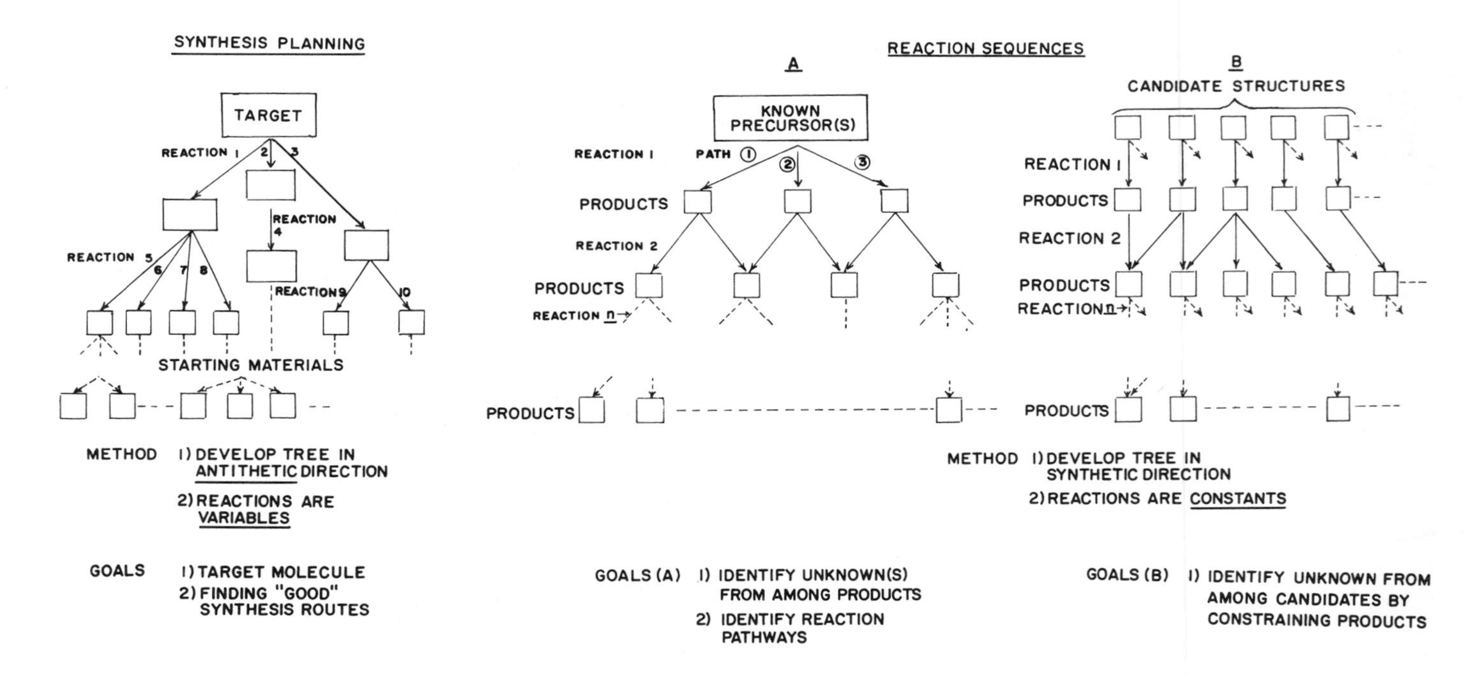

Figure 2. Contrasts between the methods and goals of development of a synthesis tree in computer-assisted synthesis and development of a reaction sequence tree in structure elucidation. We denote two applications of reaction sequences: A) conversion of known precursors into unknown compounds, a problem encountered in mechanistic studies; B) conversion of candidate structures for an unknown into a series of products, usually employed in the general problem of structure elucidation.

achieved. Reaction sequences have no predefined target molecules. The known precursors (Case A, Fig. 2) or structural candidates (Case B, Fig. 2) represent reactants. A given reaction transforms each structure which can undergo the reaction into one or more products. The products themselves may be subjected to further reaction. The goals in these reaction sequences are:

Case A) identify unknown structures in the set of products and by so doing, elucidate the reaction pathway(s).

Case B) identify the unknown structure from among the candidates by constraints applied to the products.

II) Reaction sequences operate in the synthetic rather than the retrosynthetic, or antithetic[4] direction. This difference is illustrated in Fig. 2. Each expansion of the synthesis tree represents a _set_ of reactions applied in the reverse, or antithetic direction. An actual synthesis would proceed backwards along one path. Each expansion of the CONGEN reaction tree, however, results from the application of a _single_ reaction (applied in the synthetic direction) on one or more starting materials.

III) Expansion of the synthesis tree is controlled by constraints on the reactions which are applied. Expansion of the reaction sequence tree in CONGEN is controlled by constraints on structures, _i.e._, products.

Differences (II) and (III) are reflections of the fact that synthesis programs use reactions as variables. Several reactions from a library of possible reactions may apply to any target. We consider reactions as constants. A reaction is defined (see Methods) and applied to a list of structures. The products at any level are obtained from structures at the preceding level through one or more applications of that reaction. Although many reactions may be applied to a given list of structures, leading to branches in the reaction sequence tree (see Methods) our basic task is the exhaustive exploration or evaluation, not of reactions, but of structural possibilities.

Relationship of Reaction Sequences to Structure Elucidation.

In the course of determination of the structure of an unknown compound, reactions may be carried out on the unknown to gather additional structural information. In many cases such new information can be expressed directly as constraints on the possible structures for the unknown. For example, if the base catalyzed exchange of enolizable hydrogen atoms with deuterium atoms yields a new compound whose molecular weight is three amu greater than the unknown, then the candidate structures for the unknown can be tested directly for the presence of three hydrogens exchangeable under these conditions without considering the structure of the transformed material. In fact, one criterion for useful chemical transforms designed to yield new structural information is that observations on the resulting products be easily translated back to the starting structural possibilities.

There is, however, an important class of reactions in which the translation of observations on the products into direct constraints on the structural possibilities is difficult if not impossible. In these cases it is essential to consider the application of the reaction to each structural candidate and the relationship of these candidates to their respective products. The most common examples of this class are reactions in which a given product or set of products may be obtained from different candidate structures for the unknown. Or, stated slightly differently, the class of reactions in which there is more than one way for a given product or set of products to undergo the reverse, or antithetic reaction. The following are some simple, but illustrative examples.

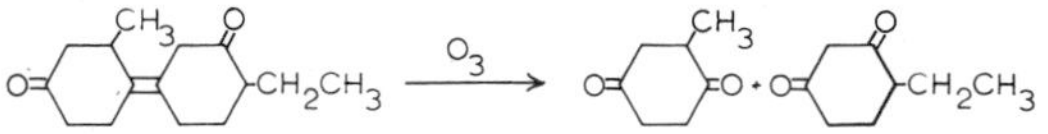

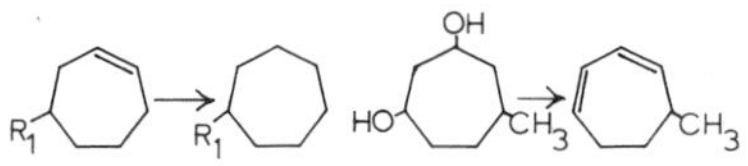

In the absence of additional information concerning the original relationships of the alkyl groups in 1 to the double bond and keto groups of 1, there are four ways of reassembling 2 and 3 (four ways of carrying out the antithetic reaction). This is an example where the functional group which was introduced is a group already present in the molecule. Thus, there is ambiguity in referring data measured on products back to the starting structures. Similar problems arise in any fragmentation reaction where more than two fragments are produced, the worst case being mass spectral fragmentations, where the fragment ions can be reassembled in many consistent ways. Transformations exemplified by 4 - 5 and 6 - 7 represent removal or introduction of multiple bonds where there is ambiguity, based on the structures of 5 and 7, concerning the structures of 4 and 6, respectively. The problem becomes rapidly more complex if a sequence of reactions is used on a set of candidate structures. Keeping the data and structural possibilities organized is a difficult job to do manually.

Other important members of this class of reactions include the cyclizations and rearrangements mentioned previously with respect to mechanistic reactions. Because these reactions generally have many ways to occur, in perhaps several consecutive steps, they are not normally used to help solve an unknown structure. Rather, such reactions are carried out on known materials and the problem is to determine the structures of observed products based at least in part on knowledge of the reaction itself. Carbonium ion rearrangements[5] and cyclization reactions such as cyclization of squalene epoxide and congeners to lanosterol and related compounds[6] are two important examples.

We designed the reaction sequence capabilities of CONGEN to make it simple for a user of the program to carry out reactions in either the general or mechanistic category of application. For the general category, measurements made on products of a reaction can be used directly to test the products without the necessity for translation of each observation back to the starting materials. These tests have direct effects on the immediate precursors of the products and eventually on the candidate structures in a multi-step sequence; removal of one product can result in eliminating whole branches of the reaction tree (see

Methods section). The complexities of dealing with many structural candidates and several reactions and associated products are handled by the internal bookkeeping of the program. In the case of mechanistic studies, the reaction capabilities are important in defining the alternative structures which can arise at each step of cyclization or rearrangement. This guarantees that all plausible products will be considered in deciding the outcome of such reactions.

Note that the ambiguities of relating observed products to starting materials are not removed by using a computer program. The advantage of using a program is that all alternatives will be systematically considered. It is easy to hypothesize one particular structure which obeys all observed data; the program provides a straightforward support or denial of such an hypothesis.

METHODS

Reaction Definition.

The initial step in carrying out a reaction on a structure or group of structures is to define the reaction and all its characteristics. This includes definition of a) the reaction site, or local environment of a molecule which will be affected by the reaction; b) the transform, or the modifications of the reaction site which yield the product(s); and c) constraints on the reaction site, or features of the local or remote environment which are either necessary for the reaction to occur or will prevent it from occurring.

The definition of the reaction is under the interactive control of the user of CONGEN (unless the reaction was previously defined to his/her satisfaction and input from a library of reactions). This introduces the need for a general, flexible and simple language of molecular structure in which the reactions can be expressed. We have adopted our general structure editor (EDITSTRUC[7]), together with the constraints mechanisms[8] of CONGEN, for reaction definition. Thus, the language used in CONGEN to describe reactions is an easily learned extension of the language needed to construct the original set of candidate structures.

EDITSTRUC is a graphical language for structure description. Each statement about rings, chains, branches, etc., results in construction of a graphical representation of the statement, in the form of a connection table. In this respect it differs from the ALCHEM language[9] developed by Wipke, which allows (restricted) English language statements about a reaction; statements which are subsequently compiled into an internal form used to carry out the reaction. Our graphical representation has the advantage that it can be used directly to carry out the reaction because all statements are understood by the current graph-matcher/pathfinder/cycle-finder and structure manipulation functions within CONGEN. Also, the important feature of the symmetry of the reaction can be computed from these graphical representations (see below). The full complement of the structural constraints available in CONGEN[8] can be brought to bear to describe the reaction in more detail. This includes the capability of specifying substructures of any complexity, variable length bridges or chains, arbitrary atom names or bond orders, proton distribution and ring sizes to be good or bad for the reaction. Wipke's ALCHEM language is currently more complete, because it allows specification of three dimensional properties and the use of certain Boolean connectives, relating constraints, which we do not currently have in CONGEN.

An example illustrating the interactive definition of a reaction and related constraints is presented in Fig. 3. The text typed by the user is underlined. The reaction is dehydrochlorination (8 -> 9), assumed to be carried out in basic conditions with the relatively bulky *t*-butoxide ion. For illustrative purposes, assume that the skeleton 10 is known, and only the placement of a chlorine atom at one of the methylene groups is in question (candidate structures then differ by this placement).

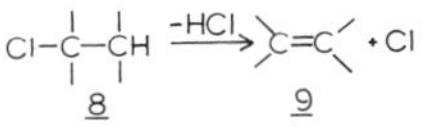

Cl–C–CH $\xrightarrow{-HCl}$ C=C + Cl

8 9 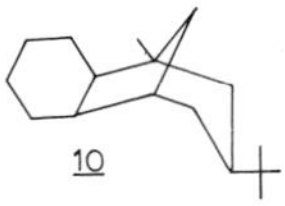 10

```
#EDITREACT
NAME: DEHYDROCHLORINATION
(NEW REACTION)

*SITE
>CHAIN 3
>ATNAME 1 CL
>HRANGE 3 1 3
>SHOW
NAME = DEHYDROCHLORINATION
ATOM  TYPE  ARTYPE NEIGHBORS HRANGE
 1     CL   NON-AR    2
 2      C   NON-AR    1 3
 3      C   NON-AR    2          1-3
>ADRAW

DEHYDROCHLORINATION: (HRANGES NOT INDICATED)

CL-C-C

>DONE


*TRANSFORM
>NDRAW

DEHYDROCHLORINATION: (HRANGES NOT INDICATED)
NON-C ATOMS: 1->CL

1-2-3

>UNJOIN 1 2
>JOIN 2 3
>ADRAW

DEHYDROCHLORINATION: (HRANGES NOT INDICATED)

CL
C=C

>DELATS 1
>DONE
```

```
*CONSTRAINTS
>BADLIST
BADLIST CONSTRAINTS
CONSTRAINT NAME:CCTBU
CONSTRAINT NAME:
-------
>DONE


*SHOW
NAME: DEHYDROCHLORINATION

SITE:
ATOM  TYPE ARTYPE NEIGHBORS HRANGE
 1     CL   NON-AR   2
 2      C   NON-AR   1 3
 3      C   NON-AR   2          1-3

DEHYDROCHLORINATION: (HRANGES NOT INDICATED)
NON-C ATOMS: 1- CL

1-2-3

TRANSFORM:
 UNJOIN 1 2
 JOIN 2 3
 DELATS 1

CONSTRAINTS:
-------
BADLIST CONSTRAINTS
  NAME
CCTBU
-------

*DONE
(DEHYDROCHLORINATION DEFINED)
(DEHYDROCHLORINATION ADDED TO THE REACTION LIST)
```

*Figure 3. An interactive session with CONGEN including definition of the reaction site (*SITE*), the reaction transform (*TRANSFORM*) and constraints on the reaction site (*CONSTRAINTS*) for the example reaction, dehydrochlorination. A summary of the complete reaction is provided by the SHOW command. User responses to CONGEN are underlined (carriage-returns terminate each command).*

Reaction Site. The reaction site represents the segment of a molecule which will be transformed. The segment includes the atoms actually involved in the reaction transformation together with any other structural features necessary for the reaction to occur. The site is defined using the appropriate EDITSTRUC commands. The reaction (8 -> 9) involves the removal of the elements of HCl from adjacent carbons. The CHAIN and ATNAME commands (Fig. 3) define the site, which is drawn for illustration (Fig. 3). The HRANGE command requires that there be from one to three hydrogen atoms on atom 3, which is obviously necessary for the elimination of HCl. (The program actually is capable of determining this itself by examination of the transform, so the HRANGE information is redundant.) The atom numbers are critical parameters for the reaction. These numbers are "sticky" in the sense that they will always be associated with the same atoms. Subsequent definition of the reaction transform itself will make explicit reference to these atom numbers.

Reaction Transform. The reaction transform is the actual series of structural modifications which, when applied to the atoms in the reaction site, yield the products. The user defines the transformation explicitly by modifications to the previously named site. The TRANSFORM command (Fig. 3) restores the actual connection table representing that site. Then, again using EDITSTRUC commands, the modifications to that site which express the reaction are defined. In the example (Fig. 3), the reaction involves loss of HCl yielding a double bond (8 -> 9), expressed as UNJOIN (break the C-Cl bond) and JOIN to form the new bond. The DELATS command deletes the chlorine atom as an inconsequential product.

Reaction Site Constraints. These constraints refer to features of the molecule (other than those in the reaction site) which affect the reaction, either positively by allowing it to occur or negatively by preventing it from occurring. These features may be in the local environment of the reaction site or may be remote as in the case of an interfering or competing functionality elsewhere in the molecule. In the example (Fig. 3), we know that this reaction will be hindered by the existing *t*-butyl group in the skeleton (10). We previously defined a substructure named, arbitrarily, CCTBU as the structure 11. Placing this substructure on BADLIST[3,8] is interpreted by CONGEN as "carry out the reaction transform at the site given by the reaction site, except when CCTBU (11) is encountered".

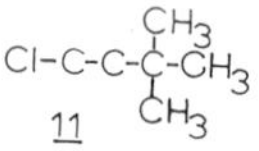

Cl-C-C-C(CH_3)(CH_3)-CH_3

11

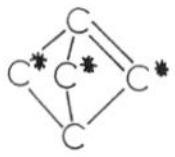

12

The SHOW command (Fig. 3) presents the user with a complete summary of the reaction in its current definition.

Carrying Out the Reaction.

Product Constraints. In many cases, certain ways of carrying out a reaction which are legal according to definitions of the reaction site, constraints and the transform yield products which are undesired. In the example (Fig. 3) we wish to avoid formation of double bonds at the bridgeheads (Bredt's rule). We supply, as a BADLIST constraint, the name of a superatom called BREDT which is previously defined as substructure 12. The starred atoms represent "linknodes" and are used to represent a path of atoms of a given length or range of lengths. The unstarred atoms in substructure 12 are the bridgeheads, the linknodes the three associated paths. The double bond in 12 is to one of the bridgehead atoms, completing an expression of the constraint.

Applying the Transform. Actual use of the reaction transform is straightforward with the exception of some features to alleviate the problems of duplication (see below). The program examines each structure in the list of structures to which the reaction is applied for the presence of the reaction site. If a site(s) is found, and the structure obeys all reaction constraints then the reaction transform is applied to the structure, once for each unique site, and a product is created for each application. Then, if the user has specified a multi-step reaction, the product molecule may be tested again for the presence of additional reaction sites and the reaction carried out again. This effectively allows us to emulate a reaction which has been carried out with a specific molar ratio of reagent to starting material; if only one mole of reagent was used, the procedure can be stopped after a single application of a transform; alternatively it can be applied to completion (exhaustively).

Consider the reaction summarized in Figure 3 and its effects on structures 13 - 15. In structure 13 the reaction site matches twice, once at C-6,7, once at C-7,8. The reaction is not carried out, however, because both reaction sites violate the undesired environment represented by 11. For 14, the reaction site matches once at C-6,12 and the reaction is carried out. But the product constraint BREDT on BADLIST (12) rejects the Bredt's rule violator 16, resulting in no products for structure 14. The reaction site fits twice in 15, at C-3,4 and C-4,5, and both fittings yield products, 17 and 18, respectively.

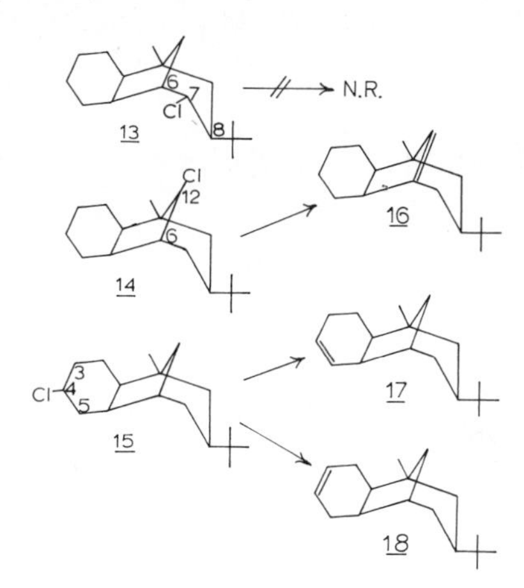

Duplication Among Products of a Reaction.

When a reaction is applied to a given list of structures, it is frequently true that some product structures occur many times in the "raw" products list. In mechanistic studies, this is the desired result because each occurrence of a product represents a unique reaction pathway (see Results and Discussion). In structure elucidation studies, though, the important information is the chemical identity of, not the pathways to, each product, and in such applications it is necessary to eliminate duplicate structures. This is not a simple matter because although structures are chemically equivalent their representations within the program may be different (e.g., the atoms may be numbered differently from one representation of a structure to the next). One method of duplicate elimination which avoids costly atom-by-atom structure comparisons between all pairs of structures involves casting each product into a standard representation ("canonical form"[10]). Duplicates then can be detected easily by direct comparison of these representations. The canonicalization process is relatively time-consuming, though, and it is desirable to explore more efficient methods of duplicate elimination wherever possible.

One type of duplication which can be detected without recourse to canonicalization is symmetry duplication, which can arise either when the reaction itself possesses some symmetry or when the starting structure is symmetrical. Our graph-matching algorithm which is responsible for locating possible fittings of the reaction site within a molecule takes no account of symmetry. For example, suppose the reaction is the addition of one mole of hydrogen to an alkene. The reaction site here is 19 and the transformation is 19 -> 20.

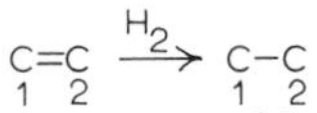

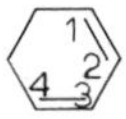

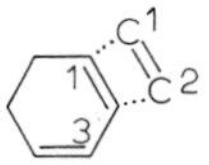

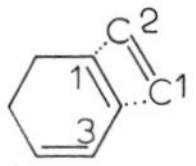

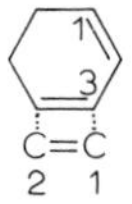

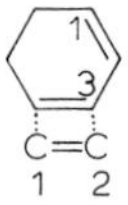

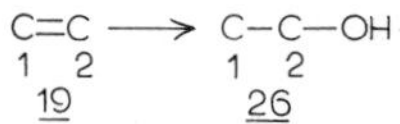

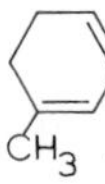

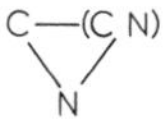

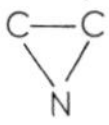

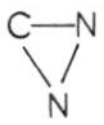

Because atoms 1 and 2 are equivalent in both the reaction site 19 and the transformed site (20), the reaction has 2-fold symmetry. If the reacting structure is 21, which itself has a two-fold symmetry plane, then the four matchings 22 - 25 all yield the same product, cyclohexene. These four matchings are members of an equivalence class determined by the symmetries of the reaction and the reacting molecule. If the reaction were unsymmetrical, say with a transform of 19 -> 26 (this would be a hydration reaction), then there would be two equivalence classes among the matchings 22 - 25, one containing 22 and 25 (each of these would lead to cyclohexen-3-ol) and one containing 23 and 24 (each yielding cyclohexen-4-ol). If the reacting molecule also had an unsymmetrical structure, say 27, then each matching would constitute a separate one-element equivalence class, and four distinct structures would result.

The general problem, then, is to eliminate all but one member of each equivalence class present in the complete set of matchings. This is a form of the so-called double coset problem of combinatorial mathematics which has been discussed previously in the context of constructive graph labeling[11]. Our solution consists of two parts. First, before the matchings are calculated, a criterion is defined for ordering any set of matchings. This criterion provides for the comparison of two matchings and, based upon the correspondence of reference numbers of the atoms in the reacting structure to reference numbers in the reaction site, defines one matching to be "smaller" than the other. Second, as each matching is obtained, the symmetry groups of both the reaction and the reacting molecule are used to form all possible symmetry images of the matching. If the matching is "smaller" than any of these symmetry images, it is kept as the representative of its equivalence class. Otherwise it is discarded as being a duplicate of some representative elsewhere in the complete set of matchings.

The symmetry of the reacting molecule is a property of its structure and can be computed prior to the matching. The symmetry of the reaction depends upon the properties (e.g., atom names, allowable ranges of hydrogens) and interconnections of the "key" atoms in the reaction site, and upon the transform modifications. Here, "key" atoms are those atoms which are actually altered by the application of the

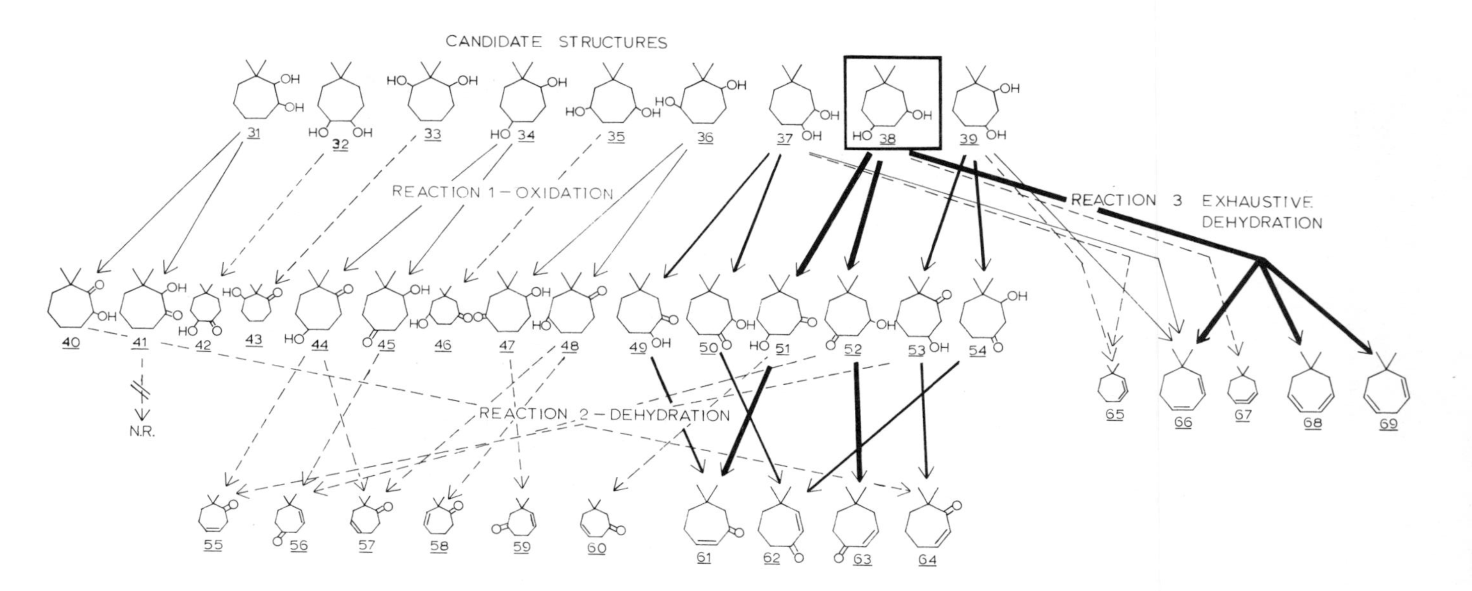

Figure 4. Development and pruning of a reaction sequence tree. Candidate structures are 31–39. Interconnecting lines and the size of a structure convey information on the fate of each candidate. Broken lines pointing to small structures mean that the product(s) and its predecessor(s) are invalid and would be removed by constraints. Regular lines mean the product(s) and its associated candidate structure remain after reaction 1; medium lines connect products and associated structures which are viable after reaction 2; and heavy lines indicate the products from the one structure, 38, which survives after all constraints are applied.

transform to the site as opposed to atoms which are necessary to, but do not directly participate in, the reaction. In some cases the properties of these key atoms, or the orders of the bonds between them, cover a range of possibilities as, for example, in the use of a special atom name which will match any non-hydrogen atom, or of a bond order of "any" which will match a bond of any multiplicity. In such cases it is not always possible to compute an overall symmetry for the reaction as a separate entity, but only the symmetry in the context of a particular matching. For example, consider the hypothetical reaction site 28 where the "polyname"[8] (C N) represents an atom which can be either C or N. This site really represents two possibilities, 29 and 30, each of which has two-fold symmetry. Only after a matching has been obtained can it be determined which of the two possibilities pertains and thus which symmetry is appropriate. In these cases it is possible at least to define, before matching, a set of possible reaction symmetries which may be applicable. Then for each matching it is necessary only to make the appropriate selection from this set, not to recompute completely the reaction symmetry.

Development, Indexing and Pruning of the Reaction Sequence Tree.

A reaction sequence may be of arbitrary complexity. A convenient representation for describing a reaction sequence is a tree structure. We illustrate the development and indexing of a reaction sequence tree in Figure 4. We assume for this example that there are nine candidate structures (31 - 39) for an unknown which is a 1,1,-cycloheptane diol, possessing no *gem*-diol functionality[12]. In the example (Fig. 4) we present the results (and their ultimate consequences) of the application of two reactions in a stepwise manner, a single-step oxidation (reaction 1) followed by a dehydration (reaction 2). A third reaction, exhaustive dehydration, is also applied to the set of candidate structures (31 - 39).

A reaction sequence tree has several important features. Branching of the tree occurs whenever more than one reaction is applied to a single set of structures (or products), *e.g.*, reactions 1 and 3, Fig. 4. There is not necessarily a one-to-one correspondence of structures to products. First, a given reaction may produce more than one product from a given structure, either: 1) because the reaction site applies more than once to the structure (*e.g.*, two products are produced from 31, 34 and 36 - 39 by reaction 1); or 2) because it is a fragmentation reaction. Second, it may be possible to obtain the same product from two different structures (*e.g.*, 61, 62 and 64 are each produced from the same reaction applied to two different structures, 49 and 51, 50 and 54, and 40 and 53 respectively).

It is possible to develop the complete reaction sequence tree by applying a planned series of reactions to the candidate structures before any laboratory work is actually done. In real applications, however, the tree would be developed in a stepwise manner by carrying out a reaction in the laboratory, acquiring data on the products and then turning to CONGEN to explore the implications of this information. We attempt to illustrate what is a very dynamic process with the static form of Figure 4.

Each reaction yields a set of products which are indexed by pointers to their precursors. These pointers are maintained at each step in the expansion of the tree so that information (constraints) applied to products at any level automatically results in appropriate action (pruning away undesired structures) at all levels below and above the given level.

There are several types of constraints which can be applied to structures in the reaction sequence tree. One constraint is a minimum to maximum number of products. In the laboratory[12], oxidation (reaction 1) of the unknown structure yielded two structures. Applying the oxidation to the set of candidate structures (31 - 39) yields two products from each structure except 32, 33 and 35, which are, therefore, rejected as candidates by CONGEN. (The structures which are single products of 32, 33 and 35 produced by CONGEN prior to application of the constraint are 42, 43 and 46, respectively.) With no further constraints, the remaining candidate structures are still viable.

Any of the existing structural constraints in CONGEN[8] can be applied to products at any step. In the laboratory, the two products of reaction 1 were separated and each subjected to a dehydration (reaction 2). A major component obtained from each product was an α,β-unsaturated ketone. Applying a GOODLIST[8] constraint expressing this observation, structures 55 -

60 are pruned away by CONGEN, leaving only the α, β-unsaturated ketones 61 - 64. This pruning also results in the rejection of products 41, 44, 45, 47 and 48 at the previous level because they did not yield α, β-unsaturated ketones. Rejecting these leads, in turn, to rejection of 31, 34 and 36 as candidate structures because their products of reaction 1 did not both yield an α, β-unsaturated ketone. This leaves only 37 - 39 as candidate structures.

Reactions can also be carried out exhaustively by repetitive application of the reaction until there are no more reaction sites remaining in the molecule. This is illustrated in the example by reaction 3, an exhaustive dehydration. In the laboratory, this reaction yielded three different dienes. Without further elaboration of the structures of these products, the correct structure can be assigned as 38 because 37 and 39 yield only one product (the same one, 66). In carrying out the reaction with CONGEN, 65 and 67 are rejected by a BADLIST constraint forbidding allenes. Products 66, 68 and 69 are produced from 38, the final structure.

When reactions are relatively simple and well understood, the kind of pruning described above can be used. As long as one has confidence in the reaction proceeding as defined, then one can use the predicted results as powerful constraints on the products and all interrelated structures in the tree. Otherwise, there would be no grounds for rejecting a structure. A characteristic of mechanistic studies, however, is that the general direction of the reaction is known, but in insufficient detail to rule out a multiplicity of products. Otherwise one could predict the products *a priori* and there would be no problem. In addition, of course, one begins with a single, known structure and it is nonsense to use the pruning mechanism described above. When we prune the reaction sequence tree for a mechanistic problem we prune out individual *pathways*. Rejection of a particular product in the tree prunes away all structures which point *only* to it or to which *only* it points. If a structure has another source or an alternative fate, it is retained. In the process of focusing in on the structures of unknown products of cyclizations or rearrangements we also focus in on the possible pathways of formation.

RESULTS AND DISCUSSION

A) An Example of Application of Reaction Sequences to Mechanistic Problems. There are at least two ways to apply the reaction sequence capabilities of CONGEN to mechanistic problems. Some of the procedures discussed subsequently can be carried out with current computer-assisted synthesis programs whenever a single compound represents the starting point.

One application of reaction sequences involves detailed mechanistic studies of possible rearrangements of a particular compound. If a mechanism is to be elucidated in detail, it is insufficient to know merely that one compound can convert to another. One must also know the identity of each atom involved and its fate in the reaction. This is normally followed by tagging various atoms in the starting material with isotopic or substituent labels. This requirement is translated in the computer program to the facility for retaining structures (at the same level in the tree) which are formally duplicates but which are in fact different in terms of the numbering of the atoms (see Methods section). Using this approach the fate of each atom at each step of a sequence can be traced. We think that this capability will help a user to design the best places for labelling the starting material, based on the computer's simulation of the course of a reaction. Consider, as a brief example, possible 1,2-alkyl shifts in structure 70, under constraints forbidding formation of 3 and 4 membered rings and methyl groups. Although there are only three new structures (in the absence of labelling) produced in the first step, there are eight different ways of performing the shift to yield four pairs of formally equivalent structures, two of which, 70a and 70b, are formally the same as the starting material (70) but have different numberings. Each of the three new skeletons appears as a pair of structures (71a,b, 72a,b, and 73a,b). The members of each pair could be distinguished by appropriate labelling of 70. Because there is a one-to-one correspondence between the atom numbers of the products and the starting material, 70, it is simple to visualize the course of each rearrangement.

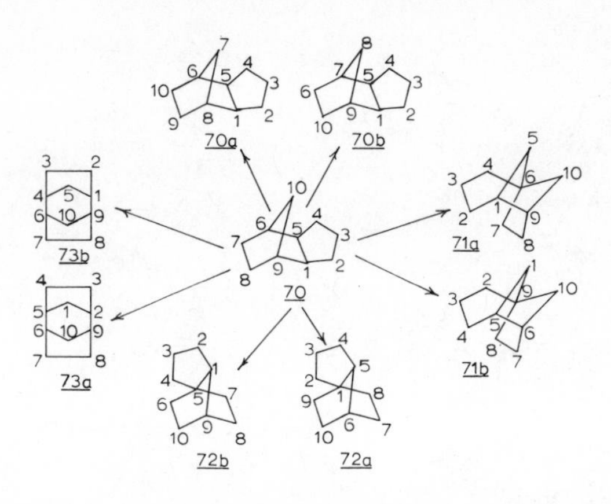

Figure 5. The eight structures, unique in terms of the numbering of their atoms, are produced by a single 1,2-alkyl shift of 70. *Transitions involving bridgehead carbonium ions were allowed, but three- and four-membered rings were forbidden.*

A second application of reaction sequences to mechanistic studies involves reactions where one needs to explore possible products and interconversion pathways, but without regard to preserving identities of atoms. An example of this type of reaction is the classic problem of the interconversion of isomers of $C_{10}H_{16}$ to adamantane. This problem has been the subject of several recent articles using computer-based approaches to help elucidate the course of various interconversions[13-15]. A characteristic of such problems is that they involve reactions which are run to completion because the reaction and associated conditions do not allow stopping the reaction after a precise number of steps have taken place. Generally, cyclizations take place until further plausible sites for cyclization are exhausted; rearrangements take place until there is no change in the ratios of products. Reaction capabilities of CONGEN can model such reactions. Cyclizations usually involve only a small number of steps, so this represents no special problem. Rearrangements, however, can proceed indefinitely if no stopping condition is specified. In the program we carry reactions to completion by applying the reaction one step at a time, stopping when no new structures, compared to all those produced previously, are encountered. Any further steps would be circular and yield no new products or pathways.

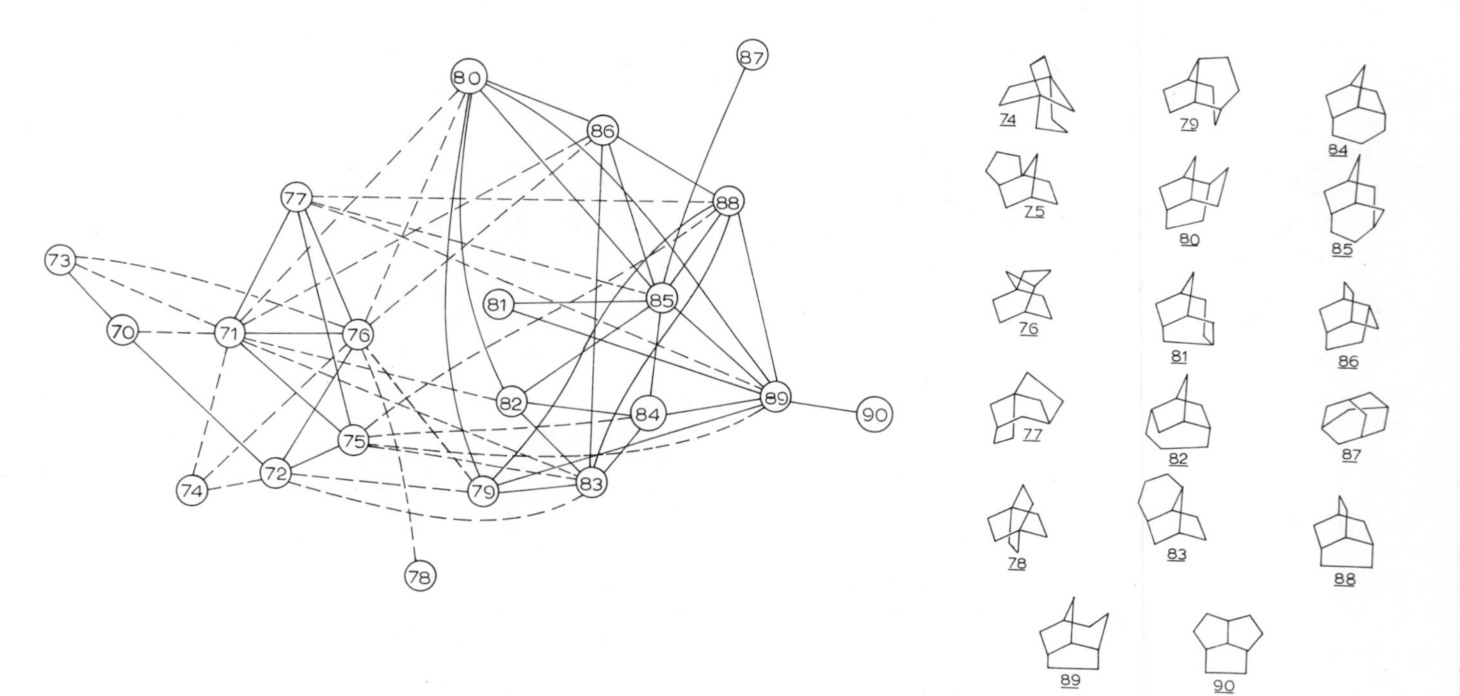

Figure 6. Complete interconversion map for $C_{10}H_{16}$ ring systems devoid of three- and four-membered rings. Pathways involving no *bridgehead carbonium ions (solid lines); pathways involving bridgehead carbonium ions (dashed lines).*

Because five structures (71, 74, 76 - 78) were missing[15] from the original set of adamantane isomers considered by Whitlock and Siefkin[13], completion of their interconversion map represents a good example for a mechanistic application. Although it is possible to do this problem as outlined above, beginning with a specific precursor and running the reaction to completion, in fact there is a much simpler way to develop the complete interconversion map. Under the structural constraints presented[13], there are 21 possible isomers[15]. Whenever the complete set of possibilities is available, the complete interconversion map can be generated by subjecting all possibilities (21 isomers) to one step of the reaction, in this case a 1,2-alkyl shift. The reverse reaction is implicit in this step and all possible pathways from one structure to others are established. The complete interconversion map is shown in Figure 6.

Earlier work[13] demonstrated that conversion of tetrahydrodicyclopentadiene (70) to adamantane (87) was not possible without invoking formation of a bridgehead carbonium ion. The existence of additional structures (71, 74, 76-78) which lie in the path of conversion of 70 to adamantane (87) meant that this question must be reinvestigated. We carried out the above rearrangement reaction under the constraints of no formation of bridgehead carbonium ions, using as definitions of bridgeheads those selected previously[13]. The results are depicted in Figure 6. We obtain a four component map if the dashed lines (pathways involving bridgehead carbonium ions) are removed. One component is 78, the second is 74, the third is 70 - 73 and 75 - 77, and the fourth is 79 - 90. Although our complete results indicate that there are alternative pathways from 70 to adamantane (87) not considered previously, the conclusion of Whitlock and Siefkin[13] that at least one bridgehead carbonium ion is required for the conversion, under the given structural constraints, is verified.

B. An Example of Application of Reaction Sequences to a Structure Elucidation Problem. The structure elucidation of coriolin[16] (whose proposed structure is 91), a sesquiterpene antibiotic, represents a problem where structural information inferred from chemical reactions played a crucial role in the tentative solution. Although it is possible in this case to translate all structural inferences derived from observations on the reaction products back to constraints on the complete set of structural possibilities, it is difficult to do using the constraints mechanism in CONGEN. In fact it is much simpler and much more intuitive, chemically, to use the reaction sequence features of CONGEN to express the reaction, obtain products, test the products with constraints and have the program automatically determine which candidate structures are plausible as a result.

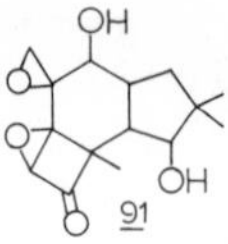

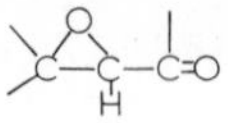

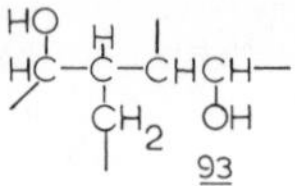

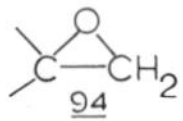

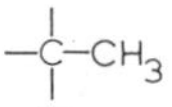

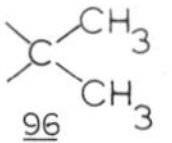

Extensive spectroscopic data revealed that coriolin has an empirical formula $C_{15}H_{20}O_5$ and is composed of five structural fragments, 92 - 96. These fragments comprise all of the atoms in the empirical formula, so that the free valences (bonds with an unspecified terminus[3]) of 92 - 96 must all be connected to other, non-hydrogen atoms.

The structural analysis of coriolin using CONGEN and reaction sequence information provides some interesting examples of the different ways both substructural and chemical inferences can be utilized to help solve the problem. If chemical experiments have already been carried out in the laboratory, then frequently some of the inferences can be used in CONGEN while constructing structures. For example, based on superatoms 92 - 96, with the constraint of no additional multiple bonds, there are more than 800 possible structures. But the chemical evidence[16] reveals that the structure possesses at least one four, one five and one six membered ring. Even though the precise environment of these rings cannot easily be specified until the reaction sequence is carried out in CONGEN, the number of each ring of each size can be used as a constraint, resulting in 56 structural candidates prior to chemical reactions. The NMR data[16] do not reveal the presence of cyclopropyl hydrogens. This constraint further reduces the number of possibilities to 52.

In the laboratory the following sequence of reactions was carried out:

CORIOLIN $\xrightarrow{H_2}$ DIHYDROCORIOLIN $\xrightarrow{LiAlH_4}$ HEXAHYDROCORIOLIN

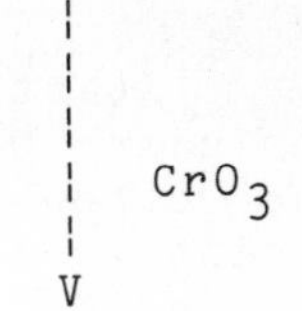

CrO_3 ↓

triketone

97: O(3) bridging CH(1)—C(2), 0-0

98: OH(3) on CH(1)—CH(2), 0-0

The first reaction reduced the ketone functionality in coriolin (91) to an alcohol. Carrying out this reaction in CONGEN yields the expected 52 products. The second reaction opened the two epoxide functionalities, yielding two new hydroxyl groups, both of which are tertiary. This allows a reaction site and transform to be defined which express this observation, (*i.e.*, 97 -> 98) a more efficient procedure than opening both epoxides both ways followed by pruning the product list. The HRANGE restriction, (*i.e.*, no hydrogens, or $H_{0->0}$), on atom 1 of 97 results in forcing the epoxide to open to a tertiary alcohol (98). The expected 52 products are obtained in CONGEN. The final reaction resulted in oxidation of the three secondary alcohol functionalities to keto groups. Spectroscopic data suggested that one of the keto groups was in a four membered ring, one in a five and one in a six membered ring. This constraint can be invoked on the original candidate structures for coriolin only by a complicated case analysis on the possible environment(s) of the original ketone functionality. As a constraint on the products of the final oxidation, however, it is a straightforward test whose ramifications in terms of structural candidates are determined automatically by CONGEN. This reaction, plus constraints, leaves 20 structural candidates, the proposed structure (91) and 19 others. We have examined these structures (automatically, using CONGEN) for the presence of naturally-occurring tricyclic sesquiterpane skeletons[17] because it was this reasoning by analogy which led to the proposal of 91 for coriolin. Structure 91 is the only one of the candidates which possesses a known skeleton.

In the group of 20 structures there are five, including 91, which obey a head-to-tail isoprene rule. The other four structures are 99 - 102. We have recently investigated[18] the scope of isomerism of terpenoid systems and find many examples in the literature where additional structural possibilities exist but where structural assignment is based on analogy with known systems. Although there may be good reasons for using analogy, no new terpenoid skeletons will be discovered this way.

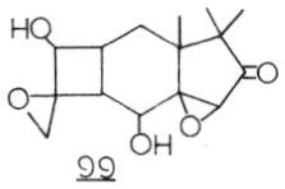

99 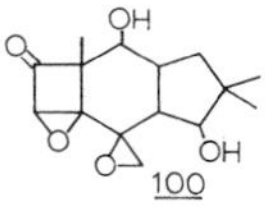 100

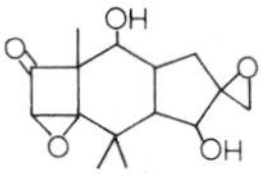

101 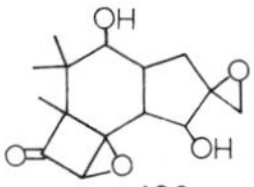 102

CONCLUSIONS

We have presented an approach which is capable of emulating many of the laboratory applications of sequences of chemical reactions. Integrated with the capabilities of the CONGEN program to suggest sets of candidate structures for an unknown compound, this approach has significantly increased the power of the program to assist chemists in solving structure elucidation problems. We have used some brief but illustrative examples to show different applications of reaction sequences to both mechanistic and structural problems. As computer programs to aid in structure elucidation develop further capabilities and become more widely available, we feel that they will be utilized exactly as other analytical tools are used. CONGEN provides a means for verifying hypotheses about unknown structures and suggesting alternatives which might otherwise be overlooked.

Our inability to utilize stereochemical information is a shortcoming, but it is not a severe problem for many applications of reaction sequences. Most chemical reactions utilized to provide additional structural, as opposed to stereochemical, information are designed to have broad application. The reactions must apply in a variety of possible situations because the environments of the functionalities involved are usually not precisely defined. In addition, detailed relative or absolute stereochemical information is seldom available until considerable detail of the structure is known; reactions cannot under these circumstances be sensitive to stereochemistry.

LITERATURE CITED

1) Part XXIII of the series "Applications of Artificial Intelligence for Chemical Inference". For Part XXII see B.G. Buchanan, D.H. Smith W.C. White, R.J. Gritter, E.A. Feigenbaum, J. Lederberg, and C. Djerassi, J. Amer. Chem. Soc., in press.
2) We wish to thank the National Institutes of Health (RR00612-05) for their generous financial support of the SUMEX computing resource (RR00785-03) on which CONGEN was developed and is made available to outside users.
3) R.E. Carhart, D.H. Smith, H. Brown and C. Djerassi, J. Amer. Chem. Soc., 97, 5755 (1975).
4) E.J. Corey and W.T. Wipke, Science, 166, 178 (1969).
5) "Carbonium Ions", Vol. I-IV, G.A. Olah and P.v.R. Schleyer, eds., Wiley Interscience, New York, N.Y. (1968-1973).
6) E.E. van Tamelen, Accts. Chem. Res., 8, 152 (1975).
7) A document describing the CONGEN program, including EDITSTRUC and all its associated structure manipulation capabilities, is available from the authors.
8) R.E. Carhart and D.H. Smith, Computers in Chemistry, in press.
9) W.T. Wipke and T.M. Dyott, J. Amer. Chem. Soc., 96, 4825 (1974).
10) H.L. Morgan, J. Chem. Doc., 5, 107 (1965).
11) L.M. Masinter, N.S. Sridharan, R.E. Carhart and D.H. Smith, J. Amer. Chem. Soc., 96, 7714 (1974).
12) This example is fictional and is used for illustrative purposes only. There are probably easier ways to distinguish the candidate structures in practice. Only structural isomers, as opposed to geometric isomers are presented.

13) H.W. Whitlock and M.W. Siefken, J. Amer. Chem. Soc., 90, 4929 (1968).
14) E.M. Engler, M. Farcasiu, A. Sevin, J.M. Cense, and P.v.R. Schleyer, J. Amer. Chem. Soc., 95, 5769 (1973).
15) R.E. Carhart, D.H. Smith, H. Brown, and N.S. Sridharan, J. Chem. Inf. Comp. Sci., 15, 124 (1975).
16) S. Takahashi, H. Iunuma, T. Takita, K. Maeda, and H. Umezawa, Tet. Lett., 4663 (1969).
17) T.K. Devon and A.I. Scott, "Handbook of Naturally Occurring Compounds. Vol. II. Terpenes," Academic Press, Inc., New York, N.Y., 1972.
18) D.H. Smith and R.E. Carhart, Tetrahedron, in press.

10

Computerized Aids to Organic Synthesis in a Pharmaceutical Research Company

D. R. EAKIN and W. A. WARR

Data Services Section, Imperial Chemical Industries Ltd., Pharmaceuticals Div., P.O. Box 25, Alderley Park, Macclesfield, Cheshire, England SK10 4TG

ICI Pharmaceuticals Division has been investigating the area of computerised aids to Organic Synthesis for many years. Our approach has been largely pragmatic and differs from other systems in two main areas. Firstly, we have assumed the research chemist is the best innovator in his specialised area - we therefore want to provide facilities which will support his intellectual capabilities, hopefully by removing some of the hit and miss aspects of reaction synthesis planning by letting computers do some of the more tedious operations. The second difference is our starting point. We felt many of the techniques already developed for the computerised manipulation of compounds should be capable of further exploitation in this area.

In the process of designing a potential drug, a pharmaceutical research chemist will have to make use of chemical information systems at various stages. He will have to find answers to such questions as:

> Is my idea novel - is anyone else working in the same or similar areas?
>
> How can I make compounds of this type?
>
> What starting materials are available to prepare compounds of this type?
>
> This compound shows activity, what compounds of similar types are available in quantities large enough to test?

Most in-house company information systems will provide some computerised aids to help chemists with these types of problem, mainly falling into one of two classes. Firstly, text searching facilities on the general literature, patent information, company reports, etc. enabling the chemist to evaluate the significance of an idea. In most systems, the chemist obtains a list of, hopefully, pertinent references which he must then pursue and evaluate. Secondly, most companies provide some substructure search facilities on their own compound files enabling the chemist to look for suitable starting materials or for compounds for biological evaluation.

We, at ICI, have tried to extend these facilities so that the chemists can readily obtain physical samples of compounds whether for synthetic purposes or not. A chemists chosen synthetic route often depends on the availability of intermediates and we have tried to expose our chemists to a wide range of available compounds which are also readily accessible.

In pharmaceutical research, chemists generally know what they want to make and often how they want to make it, and so a good supply of compounds, whether from compound stocks or commercial suppliers, becomes a valuable support facility. However, the chemist is still faced with problems in reaction synthesis. In many cases, he cannot readily find answers to questions such as:

> How was A made?
>
> How can I make compounds similar to A?
>
> What examples have we of the type of reactions that make A?

The chemist knows the product he wishes to make and must work backwards from there. He has various approaches open to him. Firstly, he can use conventional chemical information facilities such as Chemical Abstracts. Beginning with the product he wishes to make, he can examine the name and molecular formula indexes to find papers relevant to his product. By examining the abstract, he can find information on whether the compound was made by the author, and if so, how. He must, however, always look for a specific compound and may involve himself in a very time-consuming task.

His second approach is to recall a possible method of preparation and to conceive required molecular structures in terms of pathways through known or even unknown precursors. Having decided on a chosen pathway, he first examines standard textbooks to find how a certain transformation may be achieved given the constraints imposed by his particular molecule. His next step is to examine the availability of the starting material; if necessary extending his pathway back until a suitable starting material is found.

We at ICI Pharmaceuticals Division have been looking at ways in which computerised chemical information systems can be used to help chemists plan reaction syntheses more easily. We feel the chemist needs more automatic ways of finding information on:

(1) Reaction pathways.

(2) Reactant and product, particularly at the compound class level.

We felt extensions to the existing CROSSBOW facilities (1-5) could easily cope with the second factor, providing the correct data bases were available. We therefore set about examining ways of coding reaction pathways such that they could be analysed within a computer system.

THE TOTAL DESCRIPTION APPROACH

METHODOLOGY. Our first approach was to take as wide a variety of reactions as possible and code as much data about them as available, and then to draw conclusions as to how much of this data was in fact needed to satisfy user demands. We aimed to avoid applying chemical mechanistic knowledge when analysing a reaction, such that our analyses would be independent of the chemical knowledge of the encoder. To test our ideas, we chose about 700 literature-based reactions from a standard reference work (6) and recorded a wide variety of information for each reaction:

(1) Compound details for reactants and products. WLN was used so that standard CROSSBOW facilities could be used.

(2) Bibliographic details, so that the chemists could locate the originating paper for full details.

(3) Reaction conditions, including information on reagents, catalysts, solvents, time, temperature, yield and so on. A dictionary of agreed abbreviations was set-up to code this type of data in an essentially free text form.

(4) Details of the reaction site, using a variety of coding systems.

REACTION ANALYSIS. The reaction site was defined as the bonds formed or broken in a reaction and the atoms which these bonds connect. Thus, in:

$$R\text{-}\underset{\underset{O}{\|}}{C}\text{-}O\text{-}CH_3 \longrightarrow R\text{-}\underset{\underset{O}{\|}}{C}\text{-}O\text{-}H$$

it is assumed that the O⌇CH_3 bond is broken, and our supposition would be unchanged even if O^{18} labelling experiments showed that the CO-O bond were broken. We manually coded the following information for each reaction:

(a) A summary reaction classification. The classification depends upon bonds and rings formed and was based on that used at Roussel-Uclaf (7). The twelve classes were not mutually exclusive and a reaction may fall into more than one class. For example, the Fischer Indole Synthesis:

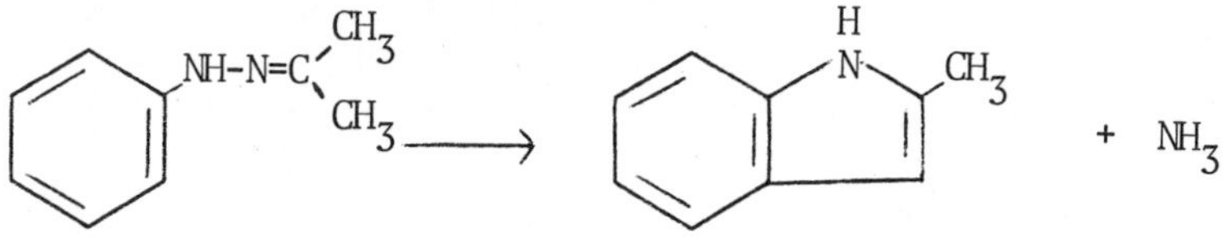

NH-N=C(CH_3)(CH_3) → (2-methylindole, N-H) + NH_3

This falls into two classes - one indicating the formation of the ring and the second the formation of the carbon-carbon atom.

(b) Details of bonds broken and formed, excluding atom-to-hydrogen bonds. No account is taken of bonds which are merely modified, for example, carbon-carbon triple bonds resulting from carbon-carbon double bonds. Consider the reaction:

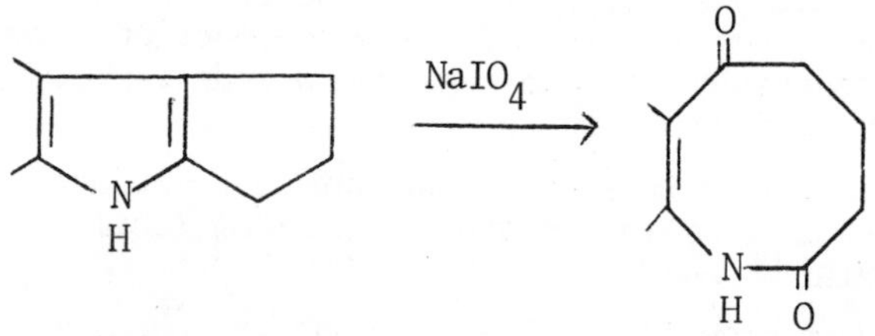

$NaIO_4$ →

The carbon-carbon double bond across the ring fusion is broken and two carbon-oxygen double bonds are formed. The bond information would be coded as:

-1C=C:+2C=O

(c) Details of individual rings broken and formed. Thus, the rings broken in the above reaction would be expressed as:

-(-55 BM

and the ring formed as:

+(-8VM EV

The coding system is based on modifications of WLN. The hyphen indicates that the ring in question is part of a larger ring system.

(d) Details of the reaction centre. Again, a system of coding was necessary to define the reaction centre, and again we used WLN with two basic modifications. Firstly, we needed to represent individual carbon atoms and we used a variety of symbols to represent the various alternatives. Secondly, we needed to indicate whether each atom was terminal, linking or branched. Using the same reaction, the reaction centre would be represented as:

DD = &OA/&OA

where "D" represents an unsaturated carbon atom at a ring fusion junction; "A" represents the ring carbon atom with an exocyclic double bond; "&" indicates the terminality of the oxygen atoms; "/" indicates the fragments are in the same molecule (8).

ASSESSMENT OF THE APPROACH. The obvious advantages of this approach lie in the large amount of data held for each reaction, and the precise statement of the reaction site. We tried to avoid the need for subjective chemical judgements either in coding or searching the file. This is not to say that chemical problems did not arise. We found that problems arose with tautomerism and mesomeric systems and with defining the reaction sites in certain re-arrangements.

However, the reaction coding involved a substantial amount of labour and we had now to ascertain the value of this, and hence, the cost-effectiveness of the approach.

The first stage was to compare the effectiveness of the system with the less labour-intensive system developed at Sheffield University (9, 10). In the latter case the recognition and generation of the reaction site is carried out by computer. All reactants and products involved in the reaction are input, and an algorithm analyses automatically the difference between reaction and product at the level of the small bond-centred fragments. The fragments which are found to be different are then re-assembled to give a skeletal reaction scheme.

Six hundred of the reactions analysed manually were subjected to the automatic analysis of reactants and products. The automatic analysis dealt with 84%, but also suffered from some faulty analyses, particularly with acylation reactions. This was partly alleviated by using an extra routine in the analysis, although this produced further defective analyses. Also sufficient information to allow precise searching was not always included. In contrast, the manual system dealt with all reactions, and did not suffer from undiscovered faulty analyses, but it was found again that the coding does not always include sufficient information to characterise the reaction.

Neither of the methods was thought ideal, consistently including all important details about the reaction. The most significant omission from a purely reaction site analysis was the position and effect of remote groups.

The second stage of the evaluation was to look at the needs of the user population. Discussions with chemists led us to believe that chemists would prefer a chemical classification and a mechanistic approach.

We therefore looked at a system based on reactant and product information coupled with a generalised reaction classification.

A SYSTEM BASED ON REACTION CLASSIFICATION

METHODOLOGY. Our second approach to reaction indexing was as down-to-earth and practical as the first was academic. We still stored the compound information on reactants and products, but reaction sites, broken bonds and so on were all discarded and a chemical/mechanistic basis was chosen for the reaction classification. The reactions studied were those which had been used by our Process Development Department for evaluation as manufacturing processes.

REACTION CLASSIFICATION. The reaction classification chosen was based on that used by the standard reference work, Organic Syntheses (11). The classes were simplified to suit the specific application. Twenty-one reaction classes were finally defined, many divided into further sub-classes. The object of the classification was to assign each stage of each reaction to only one category, unless it was a complex reaction, in which case it was assigned to as few categories as possible. Thus:

$$C_6H_5CH_2\text{-}CH(O)CH_2 \xrightarrow{R_2NH} C_6H_5CH_2CH(OH)CH_2NH_2$$

would be classed only as N-alkylation and not as a ring cleavage reaction, even though a ring is cleaved.

EVALUATING THE APPROACH. This simple approach worked reasonably well in the area to which it was applied - a small number of reactions (about 900) for which there was no available standard reference work. The system was slightly better than an equivalent published book since:

(a) The CROSSBOW facilities could be used to interrogate the reactant and product information for classes of compound.

(b) Desk-top tools could be generated to suit individual user needs, such as a molecular formula index of products, solvent indexes, etc.

We did, however, feel the computer could do more to help the chemist and set about looking for a suitable method of supplementing the reaction classification approach, bearing in mind the pitfalls of the first system.

SEMI-AUTOMATIC REACTION ANALYSIS

METHODOLOGY. We went back to think on how chemists themselves evaluate a possible synthetic route. We came to the conclusion that we needed a way to mark the reaction site information on the complete structure such that any part of the molecule can be accessed when necessary.

We started off with the basic assumption that we would code the reactant and product information in WLN, hence whatever system we devised we could make use of the substructure search facilities. We wanted to superimpose the reaction details on the reactant and product information in much the same way as a chemist would mark the reaction details on the structure diagrams indicating the reaction pathway. For example:

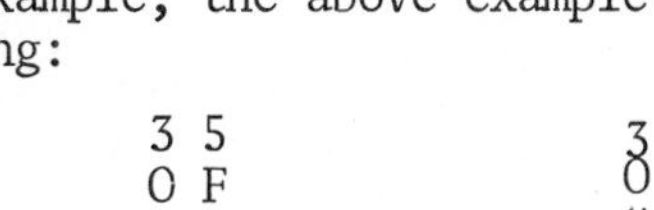

```
     O                O
     ‖                ‖
HO ≀ C-CF3  ⟶  NH2-C-CF3
```

We needed some way to indicate the reaction site and have devised a system based on the CROSSBOW connection table, which allows us to give a unique number for each node in the structure. For example, the above example gives us the following numbering:

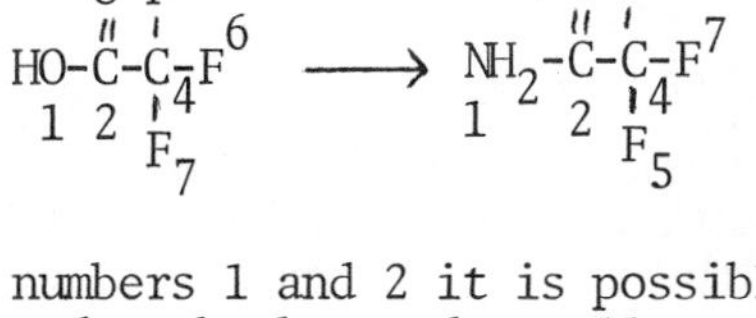

```
   3 5                    3 6
   O F                    O F
   ‖ |  6                 ‖ |  7
HO-C-C-F     ⟶    NH2-C-C-F
1  2 |4              1  2 2 |4
     F7                     F5
```

and by quoting node numbers 1 and 2 it is possible to indicate simply the bonds formed or broken and equally any rings formed or broken.

The above reaction could therefore be characterised as:

$$-1,2 \longrightarrow +1,2$$

Adaptions to conventional atom-by-atom search programs enable us to algorithmically derive the reaction centre, and we have the flexibility to make the reaction site as small or large as required by the question. Hence, in the above example, the influence of the $-CF_3$ group can be included or excluded.

Similarly, on presenting the reactions as output from a search, we can represent them as the chemists would themselves. A simple modification to our existing structure display system enables us to indicate bonds formed and broken.

EVALUATING THE SYSTEM. We are at present evaluating a system which combines the reaction classification approach with this semi-automatic site analysis. We are using the novel reactions as indicated in Index Chemicus and at the same time are evaluating the ICRS tapes as an automatic way of collecting the reaction and product information.

ICRS particularly alerts novel reactions and syntheses, and the relevant abstracts can be found by searching the tapes or by scanning Index Chemicus. The first stage of the process is to isolate all the WLNs given for the relevant abstract and to display the structures with relevant CROSSBOW connection table information.

The total reaction is examined in Index Chemicus and the reaction information added to the existing information and the appropriate class allocated. We have again used the Organic Synthesis (11) classification, but this time have left it essentially unmodified. Any relevant compounds not included in the ICRS tapes (because they did not represent novel compounds) are added to the reaction data base.

At present we are building up an index of novel reactions reported in 1975 and are pleased with the level of effort required to code any one reaction.

The necessary modifications to the atom-by-atom search program and to the structure display program have been carried out. In the case of atom-by-atom search we have included the facility to indicate the bonds in the substructure which have been formed/broken. This means that all types of substructure can be searched for and any part of that substructure may be the reaction site. Both reactant and product molecules can be searched so that the following questions can be answered:

(a) How can X be made?

(b) What happens when this bond is broken in Y?

(c) What reactions have been used to convert group X into group Z in the presence of Y?

Modifications have been made to the structure display program. Here the bonds broken or formed are marked on the structure of the total molecule. Bonds formed are marked with a circle and bonds broken with a slash mark, so imitating a chemists normal presentation.

CONCLUSIONS

Research into the methods to be used in reaction indexing and their adaptation to meet user needs has led to the development of a potentially useful reaction indexing system. It uses techniques commonly available in compound-oriented chemical information systems, with slight modifications, and yet presents data in a form readily understandable by the chemists themselves.

The resulting system should enable us to provide desk-top indexes, more up-to-date than the standard reference works and to combine these with more sophisticated computer methods. Hopefully, in this way we will improve the ways in which chemists can find chemical information relevant to chemical reaction pathways.

LITERATURE CITED

1. Hyde, E., Matthews, F.W., Thomson, L.H. and Wiswesser, W.J. - "Conversion of Wiswesser Notation to a Connectivity Matrix for Organic Compounds". J.Chem.Doc., V.7(4), p.200-204 (1967).

2. Thomson, L.H., Hyde, E. and Matthews, F.W. - "Organic Search and Display Using a Connectivity Matrix Derived from the Wiswesser Notation". J.Chem.Doc., V.7(4), p.204-207 (1967).

3. Hyde, E. and Thomson, L.H. - "Structure Display". J.Chem.Doc., V.8, p.138-146, 1968.

4. Eakin, D.R. - "The ICI CROSSBOW System" in Ash, J.E. and Hyde, E. - "Chemical Information Systems". Horwood, 1975.

5. Eakin, D.R., Hyde, E. and Palmer, G. - "The Use of Computers with Chemical Structural Information: ICI CROSSBOW System". Pesticide Sci. p.319-326, 1973.

6. Harrison, I.J. and Harrison, S. - "Compendium of Organic Synthetic Methods". Wiley, Interscience, 1971.

7. Valls, J. and Schier, O. - "Chemical Reaction Indexing" in Ash, J.E. and Hyde, E. - "Chemical Information Systems". Horwood, 1975.

8. Eakin, D.R. and Hyde, E. - "Evaluation of On-Line Techniques in a Substructure Search System" in "Computer Representation and Manipulation of Chemical Information", ed. Wipke, W.T., Heller, S.R., Feldman, R.J. and Hyde, E. Wiley.

9. Harrison, J.M. and Lynch, M.F. - "Computer Analysis of Chemical Reactions for Storage and Retrieval". Journal of the Chemical Society, p.2082-2087, 1970.

10. Seddon, J.M. - "An Evaluation of Chemical Reaction Analysis and Retrieval Systems". University of Sheffield M.Sc. Thesis, 1973.

11. Organic Synthesis. Collective Volumes, Annual Volumes. Wiley.

INDEX

INDEX

T

U

V

W

Y